Parts and Places

Parts and Places
The Structures of Spatial Representation

Roberto Casati
Achille C. Varzi

A Bradford Book
The MIT Press
Cambridge, Massachusetts
London, England

Set in Times Roman by Northeastern Graphic Services, Inc.

Printed and bound in the United States of America.

Library of Congress Cataloging-in-Publication Data

Casati, Roberto, 1961–
    Parts and places : the structures of spatial representations /
Roberto Casati, Achille C. Varzi.
      p. cm.
    "A Bradford book."
    Includes bibliographical references and index.
    ISBN 0-262-03266-X (alk. paper)
    1. Space and time. I. Varzi, Achille C. II. Title.
BD621.C35 1999
114—dc21   98-51512
                                      CIP

Contents

Acknowledgments

Space is one of the most ancient sources of intellectual puzzlement, and disciplines as diverse as pure geometry, physics, and geography have been dealing with it for many centuries. Other, younger disciplines, such as linguistics and experimental psychology, offer insights into the powers and the limits of our cognition of space. Philosophers are naturally attracted to this variety of perspectives and have been trying to reconcile them. This book is our attempt to go some way in this direction.

There are many to whom we wish to express our gratitude for helping us in this project. First, we owe much to the exchanges that we have enjoyed with friends, students, and colleagues at the Centre National de la Recherche Scientifique (CNRS) in Aix-en-Provence, Group 6059; at the Department of Philosophy of Columbia University in New York; at the Centre de Recherche in Epistemologie Appliquée (CREA) in Paris; at the Department of Geoinformation of the Technical University of Vienna; and at the Institute for System Theory and Biomedical Engineering of the Italian National Research Council (CNR) in Padova. Special thanks then go to Daniel Andler, Andrea Armeni, Paul Bloom, Luca Bonatti, Stefano Borgo, Richard Bradley, Massimiliano Carrara, Susan Carey, Dick Carter, François Clementz, Tony Cohn, John Collins, Steven Davis, Jérôme Dokic, Carola Eschenbach, Kit Fine, Andrew Frank, Marcello Frixione, Haim Gaifman, Antony Galton, Pierdaniele Giaretta, Vittorio Girotto, Nick Gotts, Nicola Guarino, Wolfgang Heydrich, Pierre Livet, Diego Marconi, David Mark, Claudio Masolo, Alain Michel, Sidney Morgenbesser, Kevin Mulligan, Ornella and Luciano Nunziante, Gloria Origgi, Elisabeth Pacherie, Mary Peterson, Jean Petitot, Fabio Pianesi, Graham Priest, Jonathan Raper, Peter Simons, Barry Smith, Roy Sorensen, Oliviero Stock, Barbara Tversky, Claude Vandeloise, Laure Vieu, and Dean Zimmerman. We would also like to thank Betty Stanton, Amy Yeager, and Paul Bethge at The MIT Press for their help and warm encouragement during the final stage of our work. Finally, we are grateful to our very special friends Rez and Alberto at Imbersago; to Holly and

Matteo in New York; and, above all, to Marianna and Friederike. They really deserve a place apart.

The preparation of the book has benefited from financial support of the Scientific and Environmental Affairs Division of the NATO Collaborative Research Grants Programme (grant CRG 960150 for the project Foundations of Spatial Representation), which we gratefully acknowledge.

Some parts of the book draw on material that have previously appeared in other places. Chapter 2 uses material from Casati and Varzi 1997b, an ancestor of which had appeared as Casati and Varzi 1995. (Some paragraphs from these papers have also been used in chapters 1 and 8.) Chapters 3 and 4 include some material from Varzi 1994, 1996b, and 1998. Chapter 5 elaborates on Varzi 1997. The first part of chapter 6 has a remote ancestor in Casati 1994. Chapter 7 consists for the most part of material from Casati and Varzi 1996. Chapter 8 contains some material from Casati 1995b and Varzi 1996a. Chapter 9 is based on Casati and Varzi 1999. Chapters 10 and 11 use material from Casati 1995a and 1999b, respectively. We express our thanks to the editors and publishers of the sources given for permission to make use of this material.

Parts and Places

1 Introduction

Thinking about space is, first and foremost, thinking about spatial things. The book is on the table; hence the table is under the book. The cake is in the tin and the tin is in the kitchen; hence the cake is in the kitchen. Sometimes we talk about things *going on* in certain places: the concert took place in the garden; the game was played at Yankee Stadium. Even when we talk about empty places—spatial regions that are not occupied by any macroscopic object and where nothing noticeable seems to be going on—we typically do so because we are planning to move things around or because we think that certain events did or could occur in certain places as opposed to others. The sofa should go right here; the accident happened right there. Spatial thinking, whether actual or hypothetical, is typically thinking about spatial entities of some sort.

One might—and some do—take this as a fundamental claim, meaning that spatial entities such as objects or events are fundamentally (cognitively, or perhaps even metaphysically) prior to space: there is no way to identify a region of space except by reference to what is or could be located or take place at that region. This was, for instance, the gist of Leibniz's contention against Newton's view that space is an individual entity in its own right which could exist even empty, with no inhabitants. At the same time, it is also worth investigating how far we can go in our understanding of spatial thinking without taking issue on such matters. Let us acknowledge the fact that empty space has little use *per se* in the ordinary representation of our environment, i.e., the representation implicit in our everyday interaction with our spatial environment. What is the meaning of this fact for a theory of spatial representation? How does it affect the construction of a general model of our spatial competence?

These questions have both a methodological and a substantive side. On the substantive side, they call for a clarification of the relevant ontological presuppositions. A good theory of spatial representation must be combined with (if not grounded on) an account of the sorts of entity that may enter into the scope of the theory, an account of the sorts of entity that can be located or take place in space—in short, an account of what may be collected under the rubric of *spatial entities* (as opposed to purely *spatial items*—as we shall say—such as points, lines, or regions). What is their distinguishing character? What special features make them spatial entities? How are they related to one another, and exactly what is their relation to space?

On the methodological side, the issue is the definition of the basic conceptual *tools* required by a theory of spatial representation, understood as a theory of the representation of these entities. There may be some ambiguity here, due to a certain ambiguity of the term 'representation'. We may think of (1) a theory of the way a cognitive system represents its spatial environment (this representation serving the twofold purpose of organizing perceptual inputs and synthesizing behavioral outputs), or (2) a theory of the spatial structure of the environment (this structure being presupposed, if not explicitly referred to, by such typical inferences as those mentioned above: The vase is in the box; hence the box is not inside the vase). The two notions are clearly distinct. Presumably, one can go a long way in the development of a cognitive theory of type 1 without developing a formal theory of type 2, and vice versa. However, both notions share a common concern; both types of theory require an account of the geometric representation of our spatial competence before we can even start looking at the mechanisms underlying our actual performances. (This is obvious for option 2. For option 1 it is particularly true if we work within a symbolic paradigm—if we favor some sort of mental logic over mental models of thinking. For then the specificity of a spatial theory of type 1 is fundamentally constrained by the structure of the domain.) It is this common concern that we have in mind here. What are the basic tools required by a theory of this spatial competence? And how do they interact?

If spatial thinking is thinking about spatial entities, we shall say that the structure of these entities must be locked into it. For instance, we can imagine a decomposition of objects and events into their parts. The table has four legs; the take-off was the most exciting part of the flight. This decomposition is often spatially relevant—How relevant is it? How do mereological (part-whole) notions interact with truly spatial concepts, such as contact, containment, and relative distance? What are the underlying principles of this interaction? And how do such concepts and principles relate to other important tools of spatial representation, such as topology, geometry, and morphology? Some of the answers hold for all sorts of spatial entities; some might not. What relevant differences are there between, say, material objects and events? Why, for instance, do spatial boundaries seem to play a crucial role for the former but not for the latter? What (if anything) distinguishes parthood from constitution? Is the clay part of the statue? Is the wood part of the table?

These questions—and many others indeed—arise forcefully as soon as we acknowledge the legitimacy of the more substantive issues mentioned above. Our contention is that the shape of the theory of space depends dramatically on the answers one gives. Over the last few years there has been considerable progress in the direction of sophisticated theories both of type 1 and of type 2, for example as part of projects aiming at the construction of information systems and autonomous machines capable of interaction with the spatial environment. We think that at this point there is some need for a philosophical pause, and our purpose in this book is to offer some thoughts that may help to fulfill this need.

Let us add immediately that we do not aim at completeness. For one thing, there is no subjectivity, no point of view embedded in the concepts that we study. Points of view are implicit in the concepts expressed by such relational predicates as 'in front of' or 'to the left of'—sometimes centered on a perceiver, sometimes grounded in some salient asymmetry in the object. Other concepts, such as those expressed by 'part of', 'contained in', or 'connected with', are "detached" concepts. We make no claim that the cluster constituted by these latter concepts can survive in isolation from "undetached" concepts; nonetheless, our concern here is exclusively with the interconnections between detached concepts. Second, this elimination of subjectivity and other perspectival facts does not automatically force upon us a realist conception of space. We simply remain as neutral as possible as to "real space" and its properties. Maybe the cognitive system reflects a true physical structure; maybe it projects the structure. But it is the structure itself that is our concern here. Third, and most important, our use of ontological notions—such as event and (material) object—is methodological. We do not offer substantive theories of these notions, and we use them only insofar as they offer the possibility of describing a rich, pluralistic spatial ontology. In particular, we do not endorse any specific psychological account of cognitively salient units or any philosophical account of individuals.

Here, then, is an overview of what follows. We begin by laying out our views on part-whole theories, and one of our main concerns is the problem of the interaction between mereological and topological concepts. There is no doubt that certain fundamental spatial structures are of a mereological nature. But what is the relevant mereology? How are the spatial parts of an object or event spatially related to one another? Traditionally, mereology has been associated with a nominalistic stance and has been put forward as a parsimonious alternative to set theory, dispensing with all abstract entities or, better, treating all entities as individuals. However, there is no necessary internal link between mereology and the philosophical position of nominalism. We may simply think of the former as a theory concerned with the analysis of parthood relations among whatever entities are allowed into the domain of discourse (including sets and other abstract entities, if one will). This certainly fits in well with the spirit of formal, type 2 theories, but cognitive, type 1 theories may also be seen this way. In this sense mereology is ontologically neutral. It is also intuitively attractive, dealing with formal structures (namely, structures of part and whole) that belong to the armamentarium not only of common sense and natural language but also of the empirical sciences. How far can we go with it? How much of the spatial universe can be grasped and described by means of purely mereological notions?

We argue that one cannot go very far. In our view (and this is a view we share with others, though we may disagree on how it can be implemented), a purely mereological outlook is too restrictive. At the very least, one needs to integrate it with concepts and principles of a topological nature. Some arguments to this effect are put forward in chapter 2, where we also argue for the complementary claim that topology itself is too poor if taken in isolation. In chapters 3 and 4 we then consider in some detail the main available options for combining mereology and topology. In effect, this need to overcome the bounds of mereology has been handled in the literature in various ways, but two main strategies can be distinguished. The first—the one we favor—is perhaps the obvious one: if topology eludes the bounds of mereology, and if its importance is to be fully recognized, then we may *add* it to a mereological basis. From this point of view, mereology can be seen as a ground theory upon which theories of greater and greater complexity (including topology as well as, say, morphology or kinematics) can be erected with the help of additional notions and principles. The second strategy is more radical. Insofar as topology is a stronger theory than mereology, one may consider turning things around: one could start from topology right away and define mereological notions in terms of topological primitives. From this point of view, just as mereology may be seen as a natural generalization of the basic theory of identity (parthood, overlapping, and even mereological fusion subsuming singular identity as a definable special case), topology may be seen as a generalization of mereology in which the general relation of connection takes over parthood and overlap as special cases. After reviewing some relevant mereological background in chapter 3, we provide an assessment of these alternative routes (and of some further options) in chapter 4, discussing their relative merits and examining to what extent their adequacy, and more generally the demarcation between mereology and topology, depends on the ontological fauna that one is willing to countenance.

Two related topics are given closer attention in the next two chapters. One concerns the distinction between interior and tangential parts. This is addressed in chapter 5. More generally, we are concerned with the notion of a boundary, and with the many philosophical conundrums that come with it. Do boundaries deserve a place of their own in the inventory of spatial articles? If so, how are they related to the entities they are boundaries of? If not, how are we to do justice to the ordinary concept of a surface, for instance? How do we explain the difference between things that are in touch and things that are simply very close? These are questions that already arise in chapter 4 when we examine the various strategies for combining mereology and topology. In chapter 5 we take stock and examine their formal and philosophical underpinnings more closely. Our position is that boundaries are ontologically on a par with (albeit parasitic upon) extended parts. But unlike extended parts, spatial boundaries have a peculiar relation to space (just as temporal boundaries have a peculiar relation to time): they are located in space, yet do not take up any space. This turns out to be a deep philosophical distinction, which occupies us for much of what follows.

The other question, which we discuss in chapter 6, concerns the very concept of part. The views presented in chapters 3–5 treat parts as objects of quantification—full-fledged entities endowed with the same right to existence as the wholes to which they belong. But such a position need be scrutinized. It goes without saying that the mereological sum of your pen and your cigar is made up of two existing things—the pen and the cigar. But what about the left and right halves of your cigar—are these *two things* of which your cigar is constituted? There is a sense in which we want to talk about proper parts such as these in the same way in which we talk about ordinary objects. We want to quantify over them, compare them, refer to them explicitly or implicitly, and so on. (The left half is heavier than the right half. The interior is darker than the rest. The cube is symmetric.) But we also want to go out of our way to make it clear that our talk about such parts is talk about potential entities—entities that do not quite count except as parts of wholes to which they belong. How exactly can we account for the potential status of such parts? How is their potentiality related to their being *undetached* from the rest of the whole to which they belong?

Chapter 7 concludes our presentation of the main ingredients of a theory of spatial representation. Besides mereology and topology, as we have already mentioned, a fundamental ingredient is the structure of location—the formal theory of the relation between an entity and the place that it occupies or where it is located. An outline of this formal theory, and of its interaction with mereological and topological concepts, forms the bulk of chapter 7. In addition, we lay the grounds for dealing with a different sort of locative relation, namely the sort of relation that holds between two objects when one is said to be inside the other (as a fly can be inside a glass). This is examined more extensively in chapter 8.

In dealing with the basic framework constituted by mereology, topology, and the theory of location, we also encounter a number of philosophical questions concerning the modal status of spatial facts. Are parts (or some parts) essential to the wholes they belong to? Are boundaries essential? And are the spatial relations among the parts and among parts and boundaries essential to the identity of the whole? We approach these issues in chapter 9. We consider, among other things, whether an object includes its place among its parts and whether it is essential for an object to be located where it is located—some novel extensions of the usual way of viewing these matters.

Chapter 10 focuses on events and their relation to space. Throughout the book we use events, as well as more ephemeral entities such as holes and shadows, as dummies for spatial entities of a different sort than material objects. Their consideration, we believe, helps disclose conceptual distinctions that might be left in the dark if one focused only on paradigmatic cases such as spatial regions or material objects. For instance, events have sometimes been put forward as a counterexample to the idea that two entities of the same kind cannot share the same spatiotemporal location. If correct, this counterexample shows that the distinction between *being located at* a spatial region and *occupying* it is not without content. On the other hand, events are eminently temporal entities: they *take time*. Their spatial structure is not as clear an issue as that of other entities, and many different aspects (such as the link between temporal and spatial boundaries and the spatial relationship between an event and its participants) must be carefully scrutinized. We do not wish to delve into the nature of events here; but we think that a study of the structural analogies and disanalogies between events and objects may play an important heuristic role in the investigation of spatial concepts vis-à-vis their temporal cognates.

In chapter 11 we deal with maps. Maps are spatial objects that *represent* other spatial objects and hence have a semantics. Our purposes there are to lay down the main features of this semantics and to draw some morals that extend to the structure of spatial representation broadly understood.

In the concluding chapter, we list a number of open issues and directions for further research.

2 Spatial Entities

Let us begin to unpack our tables and chairs. Simple as they may be, ordinary objects have structure, and a great deal of this structure is spatial. Consider:

— Objects have parts, which are hierarchically structured and spatially articulated. The table has a top and four legs; these five parts are in different places and have no parts in common, and their parts are all parts of the table. They make up the table.

— Objects are made of matter, and the matter they are made of is located exactly where the objects are located. The chair is made of wood; every part of the chair is made of wood. And where you find the chair (or part of it) you also find the wood it is made of.

— An object may be a solid whole, or it may consist of two or more disconnected parts. The table is in one piece. The red bikini has two separate pieces. The broken glass is all over the floor.

In each of these cases, the spatial structure is a central ingredient in our representation (or even in our conception) of the object. And in each case this spatial structure depends crucially on the mereological structure of the object—on the arrangement of its parts and the relations that these parts bear to the whole. This is evident in the first example (the table and its legs and top). But it is plausible also in the second case: the parts of the wood that constitutes the chair are themselves parts of the chair. (We could not move the chair without also moving the wood; we could not burn the wood and save the chair.) And when it comes to solidity and connectedness, as in the third case, we can again point to some relevant parts and explain the property of the object by reference to certain relations (of contact, for instance) among these parts.

Not only in description but also in reasoning about space we deploy a considerable mereological apparatus:

— The kitchen is part of the house. Therefore, the space taken up by the kitchen is part of the space taken up by the house.

— The table is in the kitchen and the kitchen is part of the house. Therefore, the table is in the house.

— Only part of the table is in the kitchen. Therefore, the table is only partly in the kitchen (though it may be entirely in the house).

If we proceed to a more abstract inquiry into the structure of space, we see that spatial regions are themselves mereologically organized. Indeed, in the case of spatial regions there is no difference between spatial inclusion and parthood, or between spatial intersection and mereological overlap (sharing of parts). If a region of space is included in another, it is part of that region; if two regions spatially intersect, they have a part in common.

All of this seems to suggest a simple and tempting line of thought: to uncover the spatial structure of an object (or at least a good approximation of it), look at its mereological structure. More generally, use the concepts and principles of mereology as concepts and principles for the theory of spatial representation.

Unfortunately, things are not as straightforward as this might sound. Even if the spatial structure of an object depends crucially on the relations among its parts, it does not follow that mereology—the theory of *parthood*—affords the right way to investigate such relations. Relations among parts are not necessarily parthood relations. Hence, a purely mereological outlook may be utterly inadequate for the purpose of spatial representation, and it may be necessary to integrate mereology with concepts and principles of various other sorts. In this chapter we argue that some such concepts and principles are of a topological nature. The move from mereology to mereotopology—as we shall say—is a crucial first step in the direction of a good theory of spatial representation.

2.1 Parthood and Wholeness

Consider again the opposition between a solid whole, such as a table or a cup, and a scattered object made up of several disconnected parts, such as a bikini, a broken glass, or a set of suitcases. It is true that this opposition can be explained in terms of a difference in the part-whole structures of these entities: the parts are arranged differently in the two cases. But precisely here lies the inadequacy of mereology. Strictly speaking, mereology is about parthood, hence about a relational property. By contrast, wholeness is a monadic property. And in spite of a natural and rather widespread tendency to present mereology as a theory of parts *and* wholes, the latter notion cannot be explained in terms of the former.

The grounds for this claim cannot be made fully clear except with respect to an explicit mereological framework. This is a task we attend to in the next two chapters. However, we may illustrate the underlying intuition with an example. What exactly is the difference between the cup and the broken glass? What is it that makes the cup one thing, as opposed to the many pieces of the broken glass? The difference cannot be a purely mereological one. Mereologically, for every whole there is a set of parts, and to every set of parts (that is, every arbitrary collection of objects) there may in principle correspond a complete whole, viz. their mereological sum or fusion. One could argue that not every sum is legitimate—that not every sum is a good whole. But there is no way, mereologically, to draw a distinction here; there is no way to rule out "bad" wholes consisting of scattered or ill-assorted entities (the whole consisting of the pieces of the broken glass, or the whole consisting of the broken glass, the unbroken cup, and your favorite Chinese restaurant) by thinking exclusively in terms of parthood. If we allow for the possibility of scattered entities, then we seem to lose the possibility of discriminating them from integral, connected wholes. Yet we cannot just keep integral wholes without some means of discriminating them from wholes that come in pieces.

Alfred North Whitehead's early attempts to characterize his ontology of events, as presented at length in the *Enquiry* (1919) and in *The Concept of Nature* (1920), exemplify this sort of difficulty. The mereological system underlying Whitehead's ontology was not meant to admit of arbitrary wholes, but only of wholes made up of parts "joined" or connected to one another. (For Whitehead there exist no disconnected events such as, say, the event consisting of Caesar's crossing of the Rubicon and Brutus's stabbing of Caesar.) The relevant connection relation is defined as follows[1]:

(2.1) x is connected with y if and only if there exists some z that overlaps both x and y, and that has no part that overlaps neither x nor y.

Although Whitehead was thinking of events, the intended working of this definition is easily exemplified with respect to a domain of spatial entities.

Take two bricks, and break one in half (figure 2.1). What spatial properties does the solid brick have that are missing in the broken one? What spatial properties make the two halves x and y connected in the former case but not in the latter? Well, in the former case it is easy to find parts or regions, such as z, that overlap both halves without outgrowing the whole—i.e., parts that lie entirely *within* the brick. By contrast, in the latter case it would seem that every z overlapping both x and y will also overlap their complement—i.e., the entity that surrounds the scattered whole consisting of x and y. In the latter case, every z overlapping both x and y will have to "bridge" them somehow. Not so in the former.

For Whitehead, this is all one needs to rule out scattered wholes systematically. In his own words (and with reference to his ontology of events):

Only certain pairs of events have this property. In general any event containing two events also contains parts which are separated from both events. (1919: 76)

Unfortunately, this assessment is ultimately defective. Although (2.1) does express a necessary condition for two entities x and y to be connected, it falls short of capturing a sufficient condition. And the reason is simple. There just is nothing to guarantee that the thing overlaying the two entities in question be in one piece. Figure 2.2 illustrates a simple counterexample: here z satisfies the conditions stated in (2.1), yet x and y are disconnected because z itself is disconnected. Of course, Whitehead would not allow for an entity such as z, precisely because it is not in one piece. But this is plain circularity. In other words, the account works if the general assumption is made that only self-connected entities can inhabit the domain of discourse. *But this is no account unless we can make the assumption explicit.*

These considerations apply *mutatis mutandis* to other attempts to subsume the notion of connectedness within a bare mereological frame-

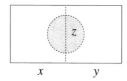 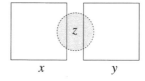

Figure 2.1
On Whitehead's account, x and y are connected (left) if there is a connecting z that lies entirely within them, and disconnected (right) if there is no such z.

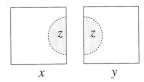

Figure 2.2
Whitehead's problem: x and y are not connected unless the overlapping piece z is itself assumed to be self-connected.[2]

work.[3] Such attempts can succeed if one goes along with Whitehead's implicit assumption that all entities are of a piece. By doing so, one would indeed support a restricted conception of wholeness in which a plurality of entities can be said to make up an integral unit just in case they have a sum (or a least an upper bound[4]). But this is no way out, for it just is not possible to make the assumption explicit by reasoning exclusively in terms of mereological relations.

2.2 The Topological Option

It is here that topology makes its entry. Connectedness is a topological relation. And if it cannot be defined in mereological terms, it must be assumed on independent grounds.

There are, of course, many other senses in which we may speak of individual integrity besides the sense of topological connectedness. This was a central intuition to Aristotle, for whom continuity was only one kind of ontological glue, along with rigidity, uniformity, and qualitative similarity (*Metaphysics* Δ, 6, 1016a). And the same intuition is to be found in Husserl's third *Logical Investigation*, which may with good reason be regarded as the first thorough formulation of a theory of parts and wholes. To Husserl, the "pregnant concept of a whole" transcends that of an arbitrary mereological aggregate and calls for an account in terms of the concept of "foundation":

By a Whole we understand a range of contents which are all covered *by a single foundation* without the help of further contents. The contents of such a range we call its parts. Talk of the *singleness of the foundation* implies that *every content is foundationally associated, whether directly or indirectly, with every content.* (1900/1901: 475, translation modified)

Various other "pregnant" concepts of whole may be distinguished in this regard. One may speak, *inter alia*, of wholes that are

— *causally unitary*: operations performed on certain parts have systematic effects on other parts (for instance, the rings of a chain form a causally connected aggregate: pulling one ring results in pulling many others)

— *functionally unitary*: a gravitational system or an electric circuit form systems that are closed under certain functional relationships (exerting gravitational force or transmitting energy, respectively)

— *teleologically unitary*: the sharing of a common goal determines a corresponding form of individual integrity (a baseball team; a party; a regiment)

— *unitary by way of dependence*: a ball projects its shadow partly on the table and partly on the carpet; it is nonetheless a single shadow.

These forms of unity are conceptually independent, and all our examples outrun spatial disunity. Consider, for instance, a world in which action at a distance is a fact. In such a world, some objects might be causally unitary without being topologically unitary. Objects with spatially scattered parts that move as one may indeed be conceptualized as unitary in spite of their scattering.

It is useful to keep this variety of concepts in mind. In what follows, however, we only make occasional references to them and we focus instead on the notion of a *topologically connected* whole as it emerges from the foregoing considerations. Overlooking microscopic discontinuities, the property or properties characterizing the spatial coherence of unitary objects are first and foremost of a topological nature.

Turning—rather boldly—the entire approach on its head, one might even suggest that topology itself provides a *more basic* and *more general* framework, subsuming mereology in its entirety. In other words, since topology eludes the bounds of mereology, one might consider reversing the order of priorities and define mereological notions in terms of suitable topological primitives. Whitehead himself, exploited this alternative route in the late formulation of his theory in *Process and Reality* (1929), following a suggestion of Theodor de Laguna (1922c). More recently this account has been worked out in great detail by Bowman Clarke (1981). On this account, the fundamental primitive is not parthood but *connec-*

tion, and all mereological notions are introduced by definition via the following:

(2.2)   *x* is part of *y* if and only if everything that is connected to *x* is also connected to *y*.

In one direction (2.2) is obvious: The leg is part of the table only if everything connected to the leg is connected to the table. But the interesting direction is the opposite one: The leg is part of the table *if* everything connected to the leg is also connected to the table.

Is this a viable account? Of course, the validity of (2.2) depends crucially on the intended interpretation of the connection relation. If we interpret it as mere overlap (sharing of parts), then (2.2) is certainly viable, but it expresses neither more nor less than a basic mereological fact: one thing is part of another just in case whatever overlaps the first overlaps the second. In fact connection is not mere overlap: the two halves of a solid brick are connected but do not overlap. So how exactly is (2.2) to be understood?

Whitehead's own account is only partially clarifying. This is because in the theory put forward in *Process and Reality* (as well as in Clarke's reformulation) the field of the connection relation is to consist exclusively of spatial or spatiotemporal regions, and not of ordinary spatial or spatiotemporal entities. In de Laguna's original suggestion, the relation of connection is conceived of as the "shadow" of the relation of physical contact—the projection of the objects onto the regions of space they may occupy. If we accept this restriction, then we may obtain a model of the theory simply by interpreting the primitive '*x* is connected to *y*' as 'region *x* and region *y* have at least one point in common'. [5] Since points are not regions, sharing of a point does not imply overlap, which therefore is not identical with (though it is a type of) connection. In other words, connecting may be done either by overlapping or by "external contact"—a distinction that makes all the difference. As Whitehead himself put it:

The possibility of this definition is [one] of the advantages gained from the adoption of Professor de Laguna's starting-point, 'extensive connection', over my original starting-point, 'extensive whole and extensive part'. (1929: 297)

Whitehead's remark is indeed no exaggeration. In Clarke's words, the distinction between overlapping and external contact "constitutes the virtue" of the account (1981: 206): external contact captures the relation of two regions just "touching" each other, and this makes it possible to distinguish between, for instance, the left and right patterns in figure 2.1. In fact, self-connectedness is now immediately defined:

(2.3) x is self-connected (i.e., in one piece) if and only if any two
       regions that make up x are connected to each other.[6]

All of this, however, applies on the hypothesis that we take our variables to range over spatial or spatiotemporal regions. If these are the only entities of our domain, then (2.2) (the definition of parthood in terms of connection) yields important conceptual achievements. The basic limits of mereology are overcome, and the result is truly a theory of parts *and* wholes. But what if we do not confine ourselves to a domain of regions?

One sort of worry is that the insistence on regions (extended entities) amounts to the exclusion of boundary elements such as points and surfaces. Whitehead, of course, had his reasons, but let us not forget that here we are looking at part-whole structures from the general perspective of spatial representation. We shall see in chapters 4 and 5 that this change of perspective introduces a number of deep philosophical conundrums.

But a second, more immediate problem arises if we interpret connection as suggested. If we really are to take an open-minded attitude toward ordinary things and events and whatever else is to be found *in* space, as we urged in chapter 1, then we do not have much choice. Either we insist on the idea that all spatial entities can be mapped onto their regions (the regions where they are located), or we maintain that a topological apparatus defined with reference to regions can be applied *holus bolus* to ordinary things and events alike.

In our view neither option seems tenable. The idea that objects are in one-one correspondence with their regions has a respectable pedigree. It was most famously defended by John Locke, who used it as a basis for a criterion of identity:

For we never finding, nor conceiving it possible, that two things of the same kind should exist in the same place at the same time, we rightly conclude that whatever exists any where at any time, excludes all of the same kind, and is there itself alone. (*Essay*, II-xxvii-1)

But Locke was thinking of entities *of a kind*, e.g., material objects. This is no guarantee that the same principle can be applied across entities of different kinds. Caesar's death (an event) took place exactly in the same region where Caesar's body (a physical object) was located. The two are distinct entities, so why should we collapse them onto one and the same thing? It is an interesting metaphysical fact, not a "convenient assumption," that these two entities shared the same spatial location. Even within the same category we can think of various sorts of immaterial or otherwise ethereal creatures—holes, shadows, ghosts, angels—for which genuine spatial interpenetration seems possible, contrary to Locke's insight.[7] For a less exotic example (from Davidson 1969), the rotation and the getting warm of a metal ball that is simultaneously rotating and getting warm are two distinct events. Neither is part of the other. Yet they occur exactly in the same spatiotemporal region, because events, unlike material objects, do not *occupy* the space at which they are located.[8]

These are reasons why we cannot accept the first option and assimilate spatial entities to their spatial regions. On the other hand, suppose we choose the second option and export our apparatus of topological notions and principles from the realm of spatial regions to the world of spatial entities. In this case the exact meaning of the connection predicate becomes crucial. What does it mean to say that two things or events share a common point? Of course we need not take Whitehead's word for that. But if we reject it, what happens to the idea that parthood can be defined in terms of connection? Rather than a starting point, it seems to us that this could at best be the outcome of a thorough analysis of the two concepts—the mereological concept of parthood and the topological concept of connection. A shadow is not part of the wall onto which it is cast, yet one could argue that everything connected to the shadow is connected to the wall. And a stone located inside a hole is not part of the hole (though the region occupied by the stone is part of the region occupied by the hole). Are they connected? If something is connected to the stone, is it connected to the hole? Or, again, suppose that every event that is topologically connected to the rotation of a ball is also connected to the ball's getting warm: would that make the ball's rotation part of its getting warm? From here on, intuitions diverge rapidly. And the notions of connection and parthood that we get by reasoning exclusively in terms of regions (along with the ensuing conceptual simplifications) just seem inadequate for dealing with the general case.

2.3 The Hole Trouble

These concerns may not be definitive. In particular, our examples of the shadow on the wall and of the stone in the hole presuppose a friendly attitude towards the ontological status of shadows and holes, which is far from pacific. (Some would just say that such "things" do not exist—after all, shadows and holes are paradigm examples of *nothings*. If so, the above question would not even arise.) We have defended our friendly attitude elsewhere.[9] Here, however, a less committing line is available to us. Let us simply use holes as well as events and ordinary objects as dummies for spatial entities that are not themselves regions of space. Our common-sense concepts (whatever their structures) of the basic entities surely are rather elastic and support a number of different interactions with spatial concepts. But the general point remains that there *might* be so many entities of different kinds in space, and these entities might have so many different relations to space, that it is hardly surprising that these differ-ences are not immediately captured by the reduction of the entities in question to "their" regions.

Be that as it may, it is apparent that the simplification introduced by reasoning only in terms of spatial regions has critical consequences if our concern is with the foundations of general-purpose theories, *even if* we take a nonrealist attitude toward holes and shadows and their likes. For the basic issue of the relationship between an entity and "its" space (the space where it is located) is then trivialized: every entity is collapsed onto its space. Moreover, this simplification yields a dull world in which impor-tant morphological features are ignored, and the question of whether holes should be treated as *bona fide* entities alongside ordinary objects is not just left in the background—it cannot even be raised. This, we main-tain, is not only a source of conceptual poverty; it may also be misleading.

Let us focus on holes. The eliminative strategy comes in a number of sorts, two of which are particularly interesting. One might want to reduce holes to regions of space (those regions, to be sure, where you expect to find the hole). Or one might want to reduce holes to some property of the objects that host them. Call the former a substantivalist and the latter an adjectivalist type of reduction. Reducing holes to regions of space would preserve the intuition that holes are individuals of a kind—hence the substantivalist flavor. But it would also rule out some of the simple, intuitive answers that we are all able to provide concerning, say, a hole's identity across time and through movement. If a hole moves, it remains the same even though it is now in a *different* region of space. (Do not think now of holes as regions of space of a new kind—for instance, mobile regions. For unless you make clear to what extent these movable regions are regions of space, you might be simply speaking of holes under another name.)

An adjectivalist reduction, on the other hand, does not consider holes as full-fledged spatial entities. Whatever we might want to say about a particular hole, the adjectivalists will translate in a peculiar idiolect of their own, where every occurrence of 'hole' is replaced by some descriptive construct that refers, not to some nonexistent hole, but rather to the host object. '*There is a hole in* this doughnut' would read, in the new idiolect, 'This doughnut *is perforated*'. Our adjectivalist, that is, behaves the same as any ordinary topologist who classifies objects' shapes by specifying the connection structure of their surfaces. To illustrate, imagine a people of two-dimensional beings—the Flatlanders—living on the surfaces of our objects and trying to find out whether their two-dimensional worlds are perforated. (They know they don't live on a flat plane, for they have discovered that if they go on a trip heading east, say, they eventually get back to the same point from the west.) They cannot conceive of a perforated object as something *with a hole*, for holes are three-dimensional. But if they are learned topologists, they could come up with a strategy (figure 2.3): 'Go on two round trips, and make sure to lay out a trail of red thread behind you. First go east and come back from the west. Then go north and come back from the south. If the two threads intersect only at the starting point, Flatland is perforated; otherwise it isn't'.[10]

There are various difficulties with this approach of which we will mention only two.

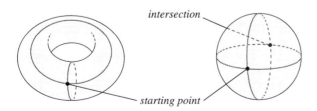

intersection

starting point

Figure 2.3
How Flatlanders can figure out the topology of their worlds.

First, the adjectivalist approach is, of course, blind to topologically insignificant holes, such as superficial hollows, grooves, and indentations. This intrinsic limit of topology shows that, in the end, the conceptual "elimination" of holes would call for a stronger theory and one that would be sensitive to all sorts of morphological features.[11] For the time being, the adjectivalist cannot even distinguish an object with a superficial hollow from an object with no hole at all. But if a theory *has no room* for the difference, its descriptive power is deficient.

Second, the adjectivalist needs a lot of adjectives to characterize the various types of hole configurations, and there is no guarantee that all types can be distinguished by reference to the topological properties of the objects. Moreover, the adjectivalist has no clear systematic way of accounting for the relationships between these adjectives. In ordinary discourse, the inference from 'The cheese has two holes in it' to 'There is at least one hole in the cheese' is straightforward. By contrast, the corresponding adjectivalist inference, from 'The cheese is doubly-perforated' to 'The cheese is at least singly-perforated' needs a special explanation.

Are there ways of avoiding this outcome without reifying holes? Well, one way would be to combine the analysis of the object's topology with that of the object's complement. But this shift—from the object to its complement—would be crucial. If one is only talking about regions, then again all is fine: there is no significant metaphysical difference between a region and its complement, and no reason to restrict oneself to one or the other. But if, as we urged, one takes an open-minded attitude towards ordinary spatial objects, then the shift to the complementary topology is ontologically significant. An object's complement is, after all, just as immaterial as a hole. The complement includes the hole. So the complementary topology of the object is, to some extent, the topology of the hole. The expressive power of the predicate 'connection' is safe. But this doesn't save us from explicit reference to immaterial entities. In short, if we care about spatial entities, then we must keep one eye on the doughnut and the other on the hole.

2.4 The Compositional Approach

We take the foregoing to imply a twofold moral. First, it appears that we need both mereology and topology as independent (though mutually related) frameworks. Mereology alone is too weak; topology alone is too strong. Second, and perhaps more important, spatial representation really goes beyond the representation of space: one must consider from the very beginning the sorts of entity that may inhabit space, and one must deal with both their mutual spatial relationships and their relationships to space. As we have already mentioned, one could even doubt the meaningfulness of representing space by itself. This issue is at the center of the controversy between spatial absolutism (the Newtonian view that space is an individual entity existing by itself, independent of whatever entities may inhabit it, and is in fact a container for the latter) and spatial relationalism (the Leibnizian view according to which space is parasitic upon, or can be construed from, relations between objects and events). But we believe we can go a long way while remaining neutral with respect to this controversy.

These delicate interconnections among mereology, topology, and spatial ontology run deep, since they manifest themselves at an elementary level. To further illustrate this point, it is worth considering one more theory that aims to analyze the mereotopological structure of ordinary objects in what appears to be a metaphysically insensitive framework: Irving Biederman's (1987) *Recognition-by-Components* (RBC) theory, possibly one of the best articulated spatial representation theories put forward in recent years. This theory—which aims to support both visual object representation and object recognition—is based on the primitive notion of a normalized cylinder, or *geon* (for *geometrical ion*), and offers a simple spatial syntax whereby every object can be viewed as composed of geonic elements.[12] The related cognitive thesis is that the human object-representation and object-recognition systems are based on a natural capacity to decompose an object into simple, abstract geometric parts. For instance, a coffee cup would decompose into a main hollow cylinder (the containing part) and a small curved cylinder (the handle) attached at both ends to the first one. (In a more recent formulation (Biederman 1990), both geons and relations among them are defined in terms of more primitive parameters, such as variation in the section size, relative size of a geon's axes with respect to its section, relative size of two geons, and vertical position of a geon at the point of junction with another. The outcome is that with only three geons one can theoretically describe some 1.4 billion distinct object shapes.)

In this theory there is no presumption that spatial entities can be dispensed with in favor of their corresponding regions. On the contrary, spatial objects are in the foreground. Their representation in terms of geons is admittedly simplistic, but a good degree of schematization seems to be appropriate if not necessary in order to provide a realistic model of our spatial competence. To see whether the table is partly in the kitchen, we don't need to give a faithful representation of the table's form, composition, and internal structure; we need only an approximation of its contour. (A sketch representing the table as composed of a top and four legs would do.) It is also for this reason, among others, that RBC has attracted some interest from computer vision and from robotics.

It is instructive, however, to look more closely at the general assumptions on which the RBC theory is based. RBC is meant to do justice to the intuition that the mereological module is crucial to object recognition. But this assumption has been questioned on empirical grounds. Cave and Kosslyn's (1993) research, for instance, indicates that a module for decomposition into parts does *not* act before, and is not a necessary condition of, object recognition. Cave and Kosslyn have shown, first, that the recognition of an object depends crucially on the proper spatial relations among the parts: when the parts are scrambled or otherwise scattered, naming times and error rates increase. Second, these results show that the mereological parsing of an object affects the object's identification "only under the most impoverished viewing conditions." This is not a disproof of the existence of a mereological module per se (for instance, the way objects are partitioned tends to be rather robust across subjects). Nevertheless, Cave and Kosslyn contend that the module need not be activated for the purpose of object recognition.

We welcome data of this sort because they dispose of the issue of object recognition in our study. In particular, the structure of the putative mereological module should be considered independent of the pressures of object recognition. The issue of the role of mereology in spatial representation remains unaffected. But let us also look, from a more abstract viewpoint, at the descriptive strength of the RBC theory. Take a flat object—say, a disk. In spite of the generative power of the notion of a normalized cylinder, it would seem that in cases like this its representational adequacy is at the limit: it seems unfair to represent a disk as a wide, flat cylinder—a collapsed geon. Of course, it might be replied that this objection matters only if our concern is with type 1 theories—with the way a cognitive system represents its spatial environment. (The fact that a certain object can be represented as a flat geon does not imply that a cognitive system actually represents it that way.) But if one is after a purely geometric theory of type 2, one could argue that this sort of artificiality is inessential. After all, for the purpose of representing the spatial structure of the objects in the environment, it does not matter what one takes a disk to be: the important thing is to keep the number of primitives to a minimum.

If, however, economy is the goal, consider again the hole problem. Take a disk with a hole—a doughnut. How is such an object to be represented by means of geons? Here the problem is twofold. On the one hand, we would again say that it is awkward to regard a doughnut (an **O**-shaped object) as consisting of two joined handles (**C**-shaped cylinders), or perhaps of a single elongated handle whose extremities are in touch. (See figure 2.4, left.) This is the type 1 misgiving. But there is also a type 2 misgiving. For how do we choose between those two different options (two geons vs. one geon)? More generally, how do we go about decomposing an object with holes in terms of its nonholed parts? There seems to be no principled way of doing that. And, of course, we wouldn't want to expand our list of primitives by adding a doughnut primitive. Otherwise, we would have to assume bitoruses—doughnuts with two holes (**8**-shaped objects)— as independent primitives. That would be necessary unless we could find a principled way within the putative RBC *cum* doughnut theory to decompose a bitorus: as doughnut plus handle (**C**-shaped geon), or as handle plus doughnut, or perhaps as a sum of two facing **3**-shaped geons (figure 2.4, right). Since the same puzzle arises also for a tritorus, and more generally for *n*-toruses, it seems that by this

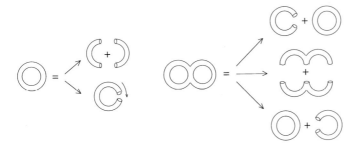

Figure 2.4
How does the RBC theory decompose a doughnut into normalized cylinders? How does it decompose a double doughnut?

pattern one would have to introduce an infinite amount of primitives—and that is undesirable also from the perspective of a type 2 theory.

It might turn out that, from a cognitive point of view, all of the alternative analyses suggested in figure 2.4 (among many others) are equally valid. That is, it might turn out that the indifference of the model correctly represents a true cognitive ambiguity. However, this would then entail that even a simple cylinder could be analyzed into two cylinders, one on top of the other. And this would seem to deprive the theory's primitive of any intuitive significance.

2.5   Negative Parts?

We can summarize the foregoing by saying that the Recognition-by-Components theory cannot simultaneously satisfy three principles that appear essential to it:

(i) The structural analysis of an object cuts at cognitively natural junctions.

(ii) The structural analysis of an object ends with geons.

(iii) Geons are (topologically equivalent to) normalized cylinders.

Once again, the problem is that the theory under consideration aims at one desideratum but neglects the others. We welcome the suggestion of investigating a spatial compositional structure that is not simply a mereology of space but of spatial entities. But we have already seen that a pure mereological prospect is not going to do all the work; *a fortiori*, one cannot go very far by reasoning exclusively in terms of such well-behaved parts as geons.

At this point one might be tempted to reconsider our earlier conclusion. Perhaps the problem is precisely that the relevant notion of part (or component) is not broad enough to do all the work. With a broader notion—not only broader than geons but also broader than the notion of part with which we started—the descriptive adequacy of mereology might be significantly enhanced. Consider, for instance, D. D. Hoffman's and W. A. Richards's suggestion that holes are parts of a kind:

[Suppose that after] two surfaces are interpenetrated one surface is pulled out of the other, leaving behind a depression, and then discarded. The depression cre-

ated in this manner has just as much motivation for being a 'part' ... as the parts
we have discussed up to this point. (1985: 84)

On this view the hole problem disappears. A doughnut is simply an object
with two main parts: a *positive* part (in the shape of a disk) and a *negative*
part (the hole). (See figure 2.5.) And since both parts can be treated as
ordinary geons, this solution would be perfectly adequate even from an
RBC perspective. (The solution is obviously generalizable to arbitrary
n-toruses. For instance, a bitorus would possess two disconnected negative parts.)

   This proposal has some attractive features of its own, which should
not be overlooked. One is that it applies to all sorts of holes, not just
topologically significant ones. The notion of a negative part does not
only account for the hole in the doughnut; it accounts for any holes you
can have in a slice of Gruyère: internal voids, perforations, and simple
hollows. This is important, for eventually we do want a unified account,
and topology cannot deliver that (as we mentioned in section 2.3). More
generally, the negative-part theory deals neatly with complementary or
dual structures, such as those constituted by grooves (notches, dents,
indentations) and ridges. A groove is a negative, intruding part, just as
a ridge is an ordinary, protruding part. There is a rather obvious reason
for the desire to treat such dual structures on a par. A natural way to
produce an indentation in a body is to act on it with another body's
protrusion; conversely, a natural way to produce a protrusion is to fuse
some material into another body's notch. We can immediately predict,
by observing the processes of fusion and indentation, that the shapes of
the notch and of the protrusion will fit perfectly into each other. Another
advantage of negative parts is that the intuition that a hole is essentially
a gap in the object (a missing something) is immediately and rather
nicely implemented. For a hole is exactly where some part of the object
could conceivably have been; and as it is *always* there where some part

Figure 2.5
Holes as negative parts.

of the object could have been, it makes no big difference if we identify it with an *actual* negative part of the object.

Let us stress that the theory of negative parts is not an eliminativist account. Wherever there is a hole, the theory recognizes a hole—a negative part. But this theory and a theory that countenances holes as entities in their own right (as we would have it) are false friends. On the hole theory, a hole is no part of its host. When you join the tips of your thumb and your index finger to form an **O**, you do not create a *new part* of yourself, however negatively you look at it. On the negative-part theory, by contrast, a hole is precisely that—a part, albeit of a somewhat special and hitherto neglected sort.

Now, this may well be a disadvantage of the hole theory. It requires a special predicate, 'hole in', logically distinct from the mereological predicate 'part of'. The two predicates are not only distinct; they are totally disjoint: holes are never part of their hosts. By contrast, if holes are treated as parts (albeit parts of a special kind) the possibility is left open that the parthood predicate be sufficient for most purposes. Presumably one would still need specific principles to the effect that, say, a negative part cannot also be a positive part of the same object. But the fundamental relations would still be part-whole relations. This is important, for conceptual economy may be very advantageous, especially from the perspective of a type 2 theory. (Ironically, this is not the perspective of Hoffman and Richards. But think, for instance, of an expert system whose task is to classify shapes. In addition to several shape primitives, one may imagine using the parthood primitive and an inversion functor that maps suitable positive parts of an object's complement onto corresponding negative parts of the object itself, and vice versa. The idea can be traced as far back as to Franz Reuleaux's 1875 descriptive *Kinematics*.)

This conceptual economy has its price, however. According to the hole theory, a doughnut is an object *with a hole*. According to the negative-part theory, a doughnut is really the *sum* of two things: a disk and a negative part. What exactly does it take to treat a doughnut that way? What kind of mereology is required? For instance, are there any constraints on how positive and negative parts can combine? Are the negative parts always included (spatially) in their positive companions? Can they occupy exactly the same region? What would be left, in that case?[13]

Furthermore, when is one allowed to speak of negative parts—what are the relevant criteria? Are holes (grooves, notches, etc.) the only sort of negative part? One may think of characterizing negative parts by reference to some intuitive normalization of the hosting wholes: a doughnut is normalized to a solid cylinder; the negative part is the "missing" cylinder in the middle. But take a sphere and cut it in half. According to one intuition, each half normalizes to the sphere itself; each half has a missing part (figure 2.6). Yet surely it would be absurd to treat a hemisphere as a whole sphere *plus* a negative part (the missing half). Of course we can regiment our intuitions and define 'normalization' more strictly. For instance, we may stipulate that every object normalizes to its convex closure (its convex hull): the doughnut would still normalize to a solid cylinder with a negative part in the middle; the hemispheres would normalize to themselves and would have no negative parts. But the solution can hardly be generalized. A champagne glass would, by that pattern, involve two large negative parts—one surrounding the stem, and one in the wine cup.

It is hard to find satisfaction in such a picture. If we don't use an explicit 'hole in' primitive, we seem to end up with a theory that is formally just as complex (owing to the need for two distinct 'part of' primitives) and ontologically more dubious (owing to the elusive nature of these mysterious missing bits). From another perspective, it is the notion of *complement* that founders conceptually. If the doughnut is really somewhat bigger than its edible part, if it also consists of a negative part, then *its* complement does not comprise the negative part. But if the mysterious negative parts are not parts of the complement—that is, if the negative parts of the doughnut are not parts of the doughnut's complement—then why are they negative? This intuition is not negotiable. And if the negativity does not lie in the complement, then why not allow for "negative" entities to begin with? Why not allow for holes?

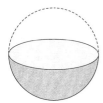

Figure 2.6
Problems with negative parts: Is a hemisphere composed of a whole sphere plus a negative half? What are the negative parts of a champagne glass?

2.6 The Need for Explicit Theories

We have again reached a general conclusion concerning the interplay of ontology, mereology, and topology. And the conclusion is that each of these three dimensions must be carefully weighed. One needs mereology because topology is mereologically unsophisticated. One needs topology because mereology is topologically blind. And one must take ontology seriously because both topology and mereology are incapable of making sense of important categorial distinctions.

Let us now take a closer look at the formal structure of this interplay by explicitly investigating the relevant primitives and the logical relationships among them. This will occupy us for the next four chapters. Not until chapter 7 will we be in a position to turn to the question of how these basic ingredients can be further combined with a theory of location proper (a theory of the relationship between a thing and its space).

3 Parthood Structures

In this and the next chapter we take a closer look at various formal accounts of the theory of parts and wholes. The guiding idea—which emerges from chapter 2—is that we can distinguish between a mereological aspect, concerned with the concept of part, and a topological aspect, concerned with the concept of a connected whole. And the main question to be examined is how these two aspects can be fruitfully combined into a unified theory (a mereotopology). We have already stated our position in this regard: neither aspect can be fully explained by the other. But we still have to see how they relate to each other. And furthermore, there are other ways of conceiving of a mereotopology, and for comparative purposes it is convenient to review them as well. In particular, we attempt a more rigorous account of the strategy mentioned in section 2.2, according to which topology would subsume mereology as a distinguished subtheory, as well as some more speculative strategies. First, however, we begin in this chapter with a general reexamination of mereology. Our purpose is both to fix notation and terminology and to set up the general theoretical background for the arguments to follow.

3.1 Preliminaries

We have been talking about mereology as a definite, unique theory, coming in the guise of a well-defined set of formal principles. This is a simplification, of course. There are very many such theories, and they do not necessarily come in similar formal guises. Indeed, the analysis of parthood relations has been a focus of philosophical investigation ever since the early days of philosophy, arguably beginning with the Presocratic atomists and continuing throughout the writings of ancient and medieval ontologists. It made its way into modern philosophy through the work of Franz Brentano and of his pupils, most notably through Husserl's third *Logical Investigation*, though it was not subjected to formal treatment until the beginning of the 20th century.[1] As we remarked in chapter 1, mereological theories have often been associated with a nominalistic stand as an ontologically parsimonious alternative to set theory, dispensing with abstract entities or (more precisely) treating all objects of quantification as individuals (entities of the lowest logical type). This is true of the two main formal theories set forth in the early years: that of Leśniewski (1916) and that of Leonard and Goodman (1940) (the latter emblematically called *Calculus of Individuals*).[2] However, there is no necessary link between mereology and the philosophical position of nominalism. As a formal theory, mereology is simply an attempt to lay down the general principles underlying the relationships between an individual and its constituent parts, just as set theory is an attempt to lay down the principles underlying the relationships between a class and its constituent members. Unlike set theory, mereology is not committed to the existence of abstract entities. The whole can be just as concrete as the parts. But mereology carries no nominalistic commitment either. The parts can be just as abstract as the whole. David Lewis's *Parts of Classes* (1991), which effectively provides a mereological analysis of the set-theoretic universe, is a good illustration of this "ontological innocence" of mereology.[3]

Historically, then, mereology has a long and complex pedigree, and our use of the term in this work involves some simplification. In fact, our concern here is mostly with contemporary formulations and developments, which grew out essentially of the two main formal theories mentioned above—Leśniewski's and Leonard and Goodman's. These are what most contemporary authors have in mind when they speak of mereology. And indeed, although those theories came in different logical guises, they are sufficiently similar to be recognized as a common basis for most subsequent developments. They are rather strong theories, however. They incorporate principles that are neither obvious nor uncontroversial, especially when it comes to explaining the notions of part and whole as these apply to spatial entities. For our purposes, therefore, it is more convenient to address the issue from a more neutral perspective and begin with an examination of some basic mereological principles.[4]

Two further caveats are in order. The first concerns the very notion of parthood that mereology is about. The word 'part' has many different meanings in ordinary language, each of which arguably corresponds to a different relation.[5] Aristotle distinguished three philosophically relevant senses in a classic passage of the *Metaphysics*:

A 'part' means (1) any kind of a division of a quantity, for what is taken from a quantity as a quantity is always called its part; two is in this sense a part of three. . . . A 'part' means (2) also the elements into which a form (not a quantity) may be divided: thus, we say that species are the parts of their genera. So, also, the elements into which any whole is divided or of which it consists, whether the whole is a form or has a form: a bronze sphere or cube has as one of its parts the bronze, which is the material in which the form is; and it also has as parts its angles. Thirdly, (3) the elements in a proposition which serves as a definition are parts of a whole; in this sense the genus is called a part of a species, though in another sense a species is part of its genus. (Δ, 1023b)

Not all of these senses fit well with the notion of parthood modeled by mereology. For instance, the relation between the bronze and the sphere cited in connection with the second sense of 'part' is sometimes distinguished from the relation that forms the focus of mereology. (Some would rather speak in this case of a relation of 'composition', or 'constitution', holding whenever two objects—in our case, the statue and the bronze—bear different relations to the same parts—their material constituents.[6]) And there are ordinary senses of 'part' that are clearly beyond the scope of any mereology, as in these examples from Cruse (1979: 29) and Winston, Chaffin, and Herrmann (1987: 426):

Changing diapers is part of being a parent.

Dating is part of adolescence.

On the other hand, Husserl began his third *Logical Investigation* with a plea for generality, which was implicitly inherited by the formal theories that followed:

We interpret the word 'part' in the *widest* sense: we may call anything a 'part' that can be distinguished 'in' an object, or, objectively phrased, that is 'present' in it. Everything is a part that is an object's real possession, not only in the sense of being a real thing, but also in the sense of being something really in something, that truly helps to make it up. (1900/1901: 437)

This too is overly broad, and one could argue that the search for the principles underlying the widest possible notion of 'part' is ultimately responsible for certain inadequacies of classical merelogical theories.[7] Yet Husserl's reference to a general notion, independent of any particular domain and free from the severe constraints of ordinary discourse, is irreproachable—particularly from our present perspective. To the extent that mereology is relevant to spatial representation, its laws may be restricted to spatially relevant cases of part-whole relationships, but these must nonetheless be grounded on sufficiently general, ontologically neutral principles. (For instance, it would be a mistake to focus exclusively on relations of inclusion among spatial regions, as we argued in chapter 2.) We therefore aim at a reasonable compromise between mereological specificity and ontological generality. This amounts to privileging structural relations of the sort exemplified by the following statements:

The tail is part of the cat.

The explosion was part of the accident.

Central Park is part of Manhattan.

The integers are part of the reals.

The relata here can be as different as material bodies, events, geographical regions, and sets of numbers. In all cases we assume there to be a structural relation of some sort between the relata, but there is no ontological restriction on the field of 'part'.

    The second caveat is related to the first. It concerns the distinction between parts and components. A component of x is a distinguished part of x, a part that is available as an individual unit regardless of its combining with other parts to make up x. (Biederman's geons, discussed in section 2.4, are components in this sense.) This distinction is to some extent reflected in the opposition between the locutions 'part of' (for parts broadly understood) and 'a part of' (for components). It is especially salient in the case of artifacts, as in the following contrast (Sanford 1993: 221):

After the explosion we found part of the motorcycle on the front lawn.

After the explosion we found a part of the motorcycle on the front lawn.

Here we are interested in the broad notion of part. Spatial partitions are not necessarily sensitive to an articulation into components. (We can localize the effects of the explosion on the motorcycle and discover that no component was left intact.) The linguistic contrast between 'part of' and 'a part of' will thus be of no consequence in the remainder of this work. Unless otherwise specified, the two expressions will be used interchangeably for the broader notion of parthood.[8]

3.2 The "Lexical" Core of a Mereological Theory

With these provisos, and barring for the moment the complications arising from the consideration of intensional factors (such as time, modalities, and counterfactuals), let us review some basic mereological principles in order of increasing strength and consequence.

A mereological theory may be viewed as a theory characterized first of all by some very basic "lexical" principles—principles setting out the cardinal features of the intended interpretation of 'part' (or of whatever other mereological predicate is taken as primitive, such as 'shares parts with' or 'overlaps'—we shall come back to this shortly). Any relation that qualifies as a candidate for parthood must satisfy such basic lexical requisites. There is, of course, a certain amount of arbitrariness in how the line is drawn between purely lexical and more substantial principles (if the opposition is meaningful at all). But most theories agree on some common ground, treating parthood as a partial ordering—a reflexive, antisymmetric, transitive relation:

Everything is part of itself.

Two distinct things cannot be part of each other.

Any part of any part of a thing is itself part of that thing.

To be sure, this characterization is not entirely uncontroversial. In particular, Rescher (1955) and several other authors have had misgivings about the transitivity of parthood.[9] Rescher writes:

In military usage, for example, persons can be parts of small units, and small units parts of larger ones; but persons are never parts of large units. Other examples are given by the various hierarchical uses of 'part'. A part (i.e., biological subunit) of a cell is not said to be a part of the organ of which that cell is a part. (1955: 10)

As we see it, these are not genuine counterexamples to transitivity. In the case of cells, the difficulty lies in the aforementioned ambiguity between 'part of' and 'a part of'. A subunit of a cell is a component (i.e., a distinguished part) of the cell, and transitivity is a general principle which is meant to apply to parts broadly understood. The military example is more to the point, yet it also trades on an ambiguity. If there is a sense of 'part' in which soldiers are not part of larger units, it is a restricted sense: a soldier is not *directly* part of a batallion—the soldier does not report to the head of the batallion. Likewise, one can argue that a handle is a functional part of a door, the door is a functional part of the house, and yet the handle is not a functional part of the house.[10] But this involves a departure from the broader notion of parthood that mereology is meant to capture. To put it differently, if the general intended interpretation of 'part' is narrowed by additional conditions (e.g., by requiring that parts make a direct contribution to the functioning of the whole), then obviously transitivity may fail. In general, if x is a ϕ-part of y and y is a ϕ-part of z, it may well be true that x is not a ϕ-part of z: the predicate modifier 'ϕ' may not distribute over parthood. But that shows the non-transitivity of 'ϕ-part' (e.g., of *direct* part, or *functional* part), not of 'part'. And within a sufficiently general framework this can easily be expressed with the help of explicit predicate modifiers.

The other two properties—reflexivity and antisymmetry—are much less controversial, though also in this regard some qualifications are in order. Concerning reflexivity, for instance, two sorts of objections may be mentioned. The first—due again to Rescher—is that

Many legitimate senses of 'part' are nonreflexive, and do not countenance saying that a whole is a part (in the sense in question) of itself. The biologists' use of 'part' for the functional sub-units of an organism is a case in point. (1955: 10)

This is of little import, though. By taking reflexivity (and antisymmetry) as constitutive of the meaning of 'part', we are just making explicit that we take identity to be a limit (improper) case of parthood. A stronger relation, whereby nothing counts as a part of itself, can obviously be defined in terms of the weaker one we are using, and so there is no loss of generality. Vice versa, one could frame the theory by taking proper parthood as a primitive instead. This is just a question of choosing a suitable primitive—a mere terminological issue.[11] (Formally, the issue therefore boils down to the previous point: a ϕ-part may not quite behave as a part *simpliciter*, where 'ϕ' is the additional condition of being distinct from the whole.) The other objection is more substantial; it says that reflexivity should fail in case we are talking about objects that are not actual (e.g., the wings of Pegasus).[12] The main reason for this restriction is that 'part', like any other predicate, should satisfy the modal "falsehood principle" of Fine (1981):

If an object does not exist in a given world, any atomic sentence to the effect that the object enjoys a certain property is false in that world.

Thus, in general (taking the relevant predicate to be 'is part of y' or 'has x as part'),

An object is part of another only if they both exist.

This is incompatible with unrestricted reflexivity. Also in this case, however, the alleged violation results from a stipulative restriction and dissolves as soon as the relevant restriction is made explicit. Obviously, unactualized parts cannot be actual parts of anything. But there is no reason to restrict mereology to the theory of the predicate 'actual part'. (There may, indeed, be good reasons to allow for "cross-world" parthood relations.) Moreover, dispensing with reflexivity introduces a disturbing dissimilarity between parthood and other distinguished predicates (such as identity, of which parthood is a generalization), unless we are also ready to make a self-identity statement '$x=x$' depend on the existence of x. We do not find this intelligible. As R. K. Meyer and Karel Lambert once put it:

It is ridiculous that from $x=x$ the logician may assert 'Caesar=Caesar', withhold comment on 'Pegasus=Pegasus' . . . , and ring up his archeological colleague with respect to 'Romulus=Romulus'. (1968: 10)

Finally, concerning the antisymmetry postulate, one may observe that this rules out "non-well-founded" mereological structures. This was pointed out by Sanford (1993: 222), who referred to Borges's Aleph as a case in point:

I saw the earth in the Aleph and in the earth the Aleph once more and the earth in the Aleph . . . (Borges 1949: 151)

We agree with van Inwagen (1993: 229) that if the only counterexamples one can think of come from literary fantasy then one may ignore them, as literary fantasy delivers no evidence in conceptual investigations. However, the idea of a non-well-founded parthood relation is not pure fantasy. In view of certain developments in non-well-founded set theory (i.e., set theory tolerating cases of self-membership and, more generally, of membership circularities),[13] one might indeed suggest building mereology on the basis of an equally less restrictive notion of 'part' that allows for parthood loops. This is particularly significant in view of the possibility of reformulating set theory in mereological terms—a possibility that is explored in the works of Harry Bunt (1985) and David Lewis (1970, 1991, 1993a). Thus, in this case there is legitimate concern that one of the principles that we are assuming to be constitutive of 'part' is in fact too restrictive. However, we are interested in mereology mainly as a general tool for spatial representation, and in this regard we don't see any use for non-well-founded parthood relations.

Now that we have clarified the main rationale behind the basic characterization of parthood as a partial ordering, it is convenient to introduce some degree of formalization before we proceed further. This avoids ambiguities (such as those involved in Rescher's objections) and facilitates comparisons and developments. We shall work within the framework of a standard first-order language with identity, supplied with a distinguished binary predicate constant, 'P', to be interpreted as the parthood relation. (The underlying logic may be taken to be a standard predicate calculus with identity.[14]). Then the above minimal requisites on parthood may be regarded as forming a theory characterized by the following proper axioms for 'P'[15]:

(P.1) Pxx *(Reflexivity)*

(P.2) $Pxy \wedge Pyx \rightarrow x=y$ *(Antisymmetry)*

(P.3) $Pxy \wedge Pyz \rightarrow Pxz.$ *(Transitivity)*

We call such a theory *Ground Mereology* (**M** for short), regarding it as the common basis of any comprehensive part-whole theory.

Given (P.1)–(P.3), one can introduce a number of additional mereological relations by exploiting the intended interpretation of 'P'. In particular, **M** supports the following intuitive definitions:

(3.1) $Oxy \ =_{df} \exists z \, (Pzx \wedge Pzy)$ (Overlap)

(3.2) $Uxy \ =_{df} \exists z \, (Pxz \wedge Pyz)$ (Underlap)

(3.3) $PPxy =_{df} Pxy \wedge \neg Pyx$ (Proper Part)

(3.4) $OXxy =_{df} Oxy \wedge \neg Pxy$ (Over-crossing)

(3.5) UXxy $=_{df}$ Uxy ∧ ¬Pyx (Under-crossing)

(3.6) POxy $=_{df}$ OXxy ∧ OXyx (Proper Overlap)

(3.7) PUxy $=_{df}$ UXxy ∧ UXyx.   (Proper Underlap)

(An intuitive model for these relations, with 'P' interpreted as spatial inclusion, is given in figure 3.1.) It is immediately verified that overlap is reflexive and symmetric, though not transitive:

(3.8) Oxx

(3.9) Oxy → Oyx.

Likewise for underlap. By contrast, proper parthood is transitive but irreflexive and asymmetric—a strict partial ordering:

(3.10) ¬PPxx

(3.11) PPxy → ¬PPyx

(3.12) PPxy ∧ PPyz → PPxz.

One could use this predicate as an alternative starting point for ground mereology (with (3.10)–(3.12) as axioms). This is because the following equivalence is provable in **M**:

(3.13) Pxy ↔ PPxy ∨ x=y

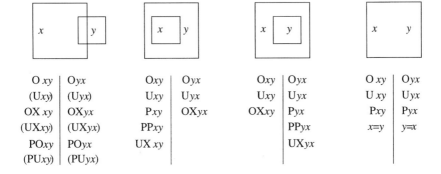

O xy	Oyx
(Uxy)	(Uyx)
OX xy	OXyx
(UXxy)	(UXyx)
POxy	POyx
(PUxy)	(PUyx)

Oxy	Oyx
Uxy	Uyx
Pxy	OXyx
PPxy	
UX xy	

Oxy	Oyx
Uxy	Uyx
OXxy	Pyx
	PPyx
	UXyx

O xy	Oyx
U xy	Uyx
Pxy	Pyx
x=y	y=x

Figure 3.1
The four basic patterns of mereological relationship. In the leftmost pattern, the relations in parenthesis hold or fail depending on whether or not there is a larger z including both x and y.

and one could therefore use the right-hand side of (3.13) to define 'P' in terms of 'PP' and '='.

On the other hand, it is worth observing that identity could itself be introduced by definition, as per the following immediate consequence of (P.2):

(3.14) $x=y \leftrightarrow Pxy \wedge Pyx$.

In that case, the theory can be formulated in a pure first-order language (without identity) as long as the characteristic axiom for identity is explicitly added to (P.1) and (P.3):

(A.I) $x=y \rightarrow (\phi x \leftrightarrow \phi y)$

(where ϕ is any formula). This would result in some conceptual economy. However, we do not pursue this route here, as we prefer to preserve a relative conceptual independence between identity and parthood. To put it differently, we take identity to be an independently clear concept, and we regard (3.14) as evidence that the parthood predicate is well-behaved in this regard.

3.3 Supplementation Principles

As we have said, the theory **M** may be viewed as embodying the lexical basis of a mereological theory. Not just any partial ordering qualifies as a part-whole relation, though, and establishing what further principles should be added to (P.1)–(P.3) is precisely the question a good mereological theory is meant to account for. These further principles are more substantive, and they are to some extent stipulative. However, some options can be identified.[16]

Generally speaking, a mereological theory may be viewed as the result of extending **M** by means of principles asserting the (conditional) existence of certain mereological items given the existence of other items. Thus, one may consider the idea that *whenever an object has a proper part, it has more than one* (i.e., that there is always some mereological difference between a whole and its proper parts). This need not be true in every model for **M**: a world with only two items, one of which is P-related to the other but not vice versa, would be a counterexample. Similarly, one may consider the idea that *there is always a mereological sum of two or more* *parts* (i.e., that for any number of objects there exists a whole that consists exactly of those objects). Again, this need not be true in a model for **M**. More generally, one may consider extending **M** by requiring that the domain of discourse be closed—on certain conditions—under various mereological operations (sum, product, difference, and possibly others). Finally, one may consider the question of whether there are any *mereological atoms* (objects with no proper parts), and also whether every object is *ultimately made up of atoms* (or on what conditions an object may be assumed to be made up of atoms). Both of these options are compatible with **M**, and the possibility of adding corresponding axioms has interesting philosophical ramifications.

Let us, then, begin with the first sort of extension. The underlying idea can take at least two distinct forms.

The simplest one consists in strengthening **M** by adding a fourth axiom to the effect that every proper part must be *supplemented* by another, disjoint part:

(P.4) $PPxy \rightarrow \exists z(Pzy \wedge \neg Ozx)$.   *(Weak Supplementation)*

We call this extension *Minimal Mereology* (**MM**). Some authors (most notably Peter Simons[17]) regard (P.4) as constitutive of the meaning of 'part' and would accordingly list it along with the lexical postulates of mereology. However, some mereotopological theories in the literature violate this principle, and it is therefore convenient to keep it separate from (P.1)–(P.3). (We are, of course, especially interested in spatial applications of mereology. Thus, we are not impressed by counterexamples to (P.4) that bring in non-extensive differences. For instance, for Brentano (1933) a soul is a proper part of a thinking soul, but there exists no extra part to make up the difference.[18])

The second way of expressing the supplementation intuition is stronger. It corresponds to the following axiom:

(P.5) $\neg Pyx \rightarrow \exists z(Pzy \wedge \neg Ozx)$.   *(Strong Supplementation)*

It is easily seen that (P.5) implies (P.4). However, the converse does not hold. For instance, consider two distinct objects *a* and *b* that have exactly the same two proper parts, say *c* and *d*. Then the corresponding instance of (P.4) is true, since each proper part counts as a supplement of the other; yet (P.5) is false, since both parts of *a* are part of (and therefore overlap) *b*, and both parts of *b* are part of (and therefore overlap) *a*. Admittedly, it is difficult to *imagine* such objects; it is difficult to draw a picture illustrating two distinct objects with the same parts, because drawing an object *is* drawing its parts. Once the parts are drawn, there is nothing left to be done to get a drawing of the whole object. But this proves only that pictures are biased towards (P.5). It does not give us reason enough to conclude that spatially extended entities must obey (P.5). Such a conclusion would once more reflect a confusion between the structure of spatial entities and that of their spatial locations—a confusion we started to unveil in chapter 2. (In the non-spatial domain, the envisaged countermodel to (P.5) is obtained by letting *a* be the ordered pair $\langle c, d \rangle$ and *b* the ordered pair $\langle d, c \rangle$ and interpreting 'P' as the relation of membership for ordered sets.)

The theory obtained by adding (P.5) to (P.1)–(P.3) is thus a proper extension of the theory of Minimal Mereology obtained by adding (P.4). We label this stronger theory *Extensional Mereology* (**EM**). The attribute 'extensional' is justified precisely by the exclusion of countermodels that, like the one just mentioned, contain distinct objects with the same proper parts. In fact, it is a theorem of **EM** that non-atomic objects with the same proper parts are identical:

(3.15) $(\exists z \mathrm{PP} z x \lor \exists z \mathrm{PP} z y) \to (\forall z (\mathrm{PP} z x \leftrightarrow \mathrm{PP} z y) \to x = y)$.

(The analogue for 'P' is already true in **M**, since P is reflexive and antisymmetric.) This is the mereological counterpart of the familiar set-theoretic extensionality principle, as it reflects the view that an object is exhaustively defined by its constituent parts, just as a set is exhaustively defined by its constituent elements. Nelson Goodman appropriately termed this mereological principle "hyper-extensionalism" (1958: 66), taking it to express the ultimate criterion for nominalism:

A class (e.g., that of the counties of Utah) is different neither from the single individual (the whole state of Utah) that exactly contains its members nor from any other class (e.g., that of acres of Utah) whose members exactly exhaust this same whole. The platonist may distinguish these entities by venturing into a new dimension of Pure Form, but the nominalist recognizes no distinction of entities without a distinction of content. (Goodman 1951: 26).

There are, of course, several *prima facie* difficulties with this idea, in spite of its naturality in the context or a purely spatial domain. One recurrent objection is that there are or may be individuals that differ exclusively with respect to the *arrangement* of their parts. Two sentences made up of the same words—'John loves Mary' and 'Mary loves John'—would be a case in point.[19] A second familiar objection is closely related (though the relation has not always been made explicit). If the equality sign in (3.15) stands for identity, and if this is a transitive relation, then the principle of mereological extensionality seems to contradict the possibility that an object may undergo mereological change without losing its identity. This has often been illustrated with reference to a puzzle that arises in connection with the mereology of continuants. A cat can survive the annihilation of its tail. A thing consisting of the cat's tail and the rest of the cat's body cannot survive the annihilation of the tail. Thus, a cat and a thing consisting of the cat's body plus its tail have different (modal) properties. Ergo they cannot be identical (even if they have the same parts).[20]

The first of these objections is easily dispensed with. Sentences made up of the same words are best described as different sentence *tokens* made up of distinct tokens of the same word *types*. There is, accordingly, no violation of extensionality. The second objection is more serious: as a criterion for individual identity, mereological extensionality is indeed very strict. At the same time, to abandon it and start distinguishing among mereologically indistinguishable entities is like opening Pandora's box, and may end up in a multiplication of entities beyond tolerance: if the cat is different from the entity tail+remainder, it must also be different from the entity head+remainder, and from the entity nose+remainder, and so on. How many entities then occupy the region occupied by the cat? To what principled criterion can we appeal to avoid this slippery slope? (Note that the slippery slope is a real possibility once we have rejected the idea that to each place there corresponds a unique entity.)

We must be careful here. There is, to begin with, an ambiguity in the descriptive terms used to designate these purportedly distinct entities. The statement that *a thing consisting of the cat's tail and the rest of its body* cannot survive the annihilation of the tail can be understood as expressing a *de dicto* modality:

For every world *w*, a thing that in *w* consists of the cat's tail and the rest of its body must have the cat's tail as a (proper) part in *w*.

Alternatively, the statement can be understood as expressing a *de re* modality:

A thing that in the actual world consists of the cat's tail and the rest of its body must have the cat's tail as a (proper) part in every world *w*.

On the *de dicto* reading, the statement is obviously unproblematic with regard to extensionality: it is a true statement (a logical truth, in effect), but it does not imply anything concerning the identity of things consisting of the cat's tail and the rest of its body *in the actual world*. On the *de re* reading, the statement is not in itself unproblematic, yet again it is not in conflict with extensionality: as far as the story goes, that thing that in the actual world consists of the cat's tail and the rest of its body may just be our old cat. In other words, our descriptive terms ('the cat' and 'a thing consisting of the cat's tail and the rest of its body') have different senses, but they may very well have the same referent. If so, then on the *de re* reading our statement is true just in case it is true that *the cat* can survive the annihilation of the tail. (Surely after the tail had been annihilated it would be inappropriate to refer to the cat as 'a thing consisting of the cat's tail and the rest of its body', just as it would be wrong to refer to the current president of the United States as 'the current US President' after the presidency was over.) We are no longer forced to distinguish between one entity and the other, so we have no compelling reason to reject extensionality.[21]

   This is not to deny that survival through mereological change is a real issue. Even if the puzzle dissolves, we are left with the fundamental question: Does the cat—that unique thing consisting of that body and that tail—indeed survive the annihilation of the tail? Arguably, the answer involves modal notions. For instance, we could consider a weaker interpretation of the equality sign in (P.2) and (3.15) as standing, not for identity, but simply for mereological coincidence *at a given time*. Identity would accordingly amount to mereological coincidence *at all times*; we could then rely on this distinction to explain away the dilemma.[22] Alternatively, we could insist that losing or gaining parts implies a change of identity: the cat is no longer the same, but a natural continuity between the thing before and after the loss of the tail justifies our referring to *them* as being one and the same thing enduring through time.[23] Here we leave the question open. We regard **EM** as the natural way of implementing the extentionalist idea. But from time to time we shall discuss the question of the adequacy of the idea itself.

3.4 Closure Principles

The second way of extending **M** corresponds to the idea that a mereological domain must be closed under various operations. Take first the quasi-Boolean operations of sum and product. (Mereological sum is sometimes called fusion.) If two things underlap, then we may assume that there is a smallest thing of which they are part—a thing that exactly and completely exhaust both. For instance, your left thumb and index finger underlap, since they are both parts of you. There are other things of which they are part—e.g., your left hand. But we may assume that there is a smallest such thing: that part of your left hand which consists exactly of your left thumb and index finger. Likewise, if two things overlap (e.g., two intersecting roads), then we may assume that there is a largest thing that is part of both (the common part at their junction). These two assumptions can be expressed by means of the following axioms, respectively:

$$(\text{P.6}) \quad Uxy \to \exists z \forall w (Owz \leftrightarrow (Owx \lor Owy)) \qquad \qquad (Sum)$$

$$(\text{P.7}) \quad Oxy \to \exists z \forall w (Pwz \leftrightarrow (Pwx \land Pwy)). \qquad \qquad (Product)$$

Call the extension of **M** obtained by adding (P.6) and (P.7) *Closure Mereology* (**CM**). The result of adding these same axioms to **MM** or **EM** yields corresponding *Minimal* or *Extensional Closure Mereologies* (**CMM** and **CEM**), respectively.

The intuitive idea behind these two axioms is best appreciated in the presence of extensionality, for in that case the entities whose conditional existence is asserted by (P.6) and (P.7) must be unique. Thus, if the language has a description operator 'ι', **CEM** supports the following definitions[24]:

$$(3.16) \quad x+y =_{\text{df}} \iota z \forall w (Owz \leftrightarrow (Owx \lor Owy))$$

$$(3.17) \quad x \times y =_{\text{df}} \iota z \forall w (Pwz \leftrightarrow (Pwx \land Pwy))$$

and (P.6) and (P.7) can be rephrased more perspicuously as

$$(\text{P.6}') \quad Uxy \to \exists z (z = x+y)$$

$$(\text{P.7}') \quad Oxy \to \exists z (z = x \times y).$$

In other words, any two underlapping things have a unique mereological sum, and any two overlapping things have a unique product. Actually the connection with extensionality is subtler. In the presence of the Weak Supplementation Principle (P.4), the product closure (P.7) implies the Strong Supplementation Principle (P.5). Thus, **CMM** turns out to be the same theory as **CEM**.

In the literature, closure merelogies are just as controversial as extensional mereologies, though for quite independent reasons. The main objection is that they involve an increase in the number of entities admitted in a mereological domain. Indeed there is no question that a **C(E)M** model may be more densely populated than a corresponding **(E)M** (or **MM**) model. However, one could argue that the ontological increase is not real—there is no further ontological commitment. After all, a product *adds* nothing, and even a sum is, in a sense, nothing over and above its constituent parts. As David Lewis put it:

> Given a prior commitment to cats, say, a commitment to cat-fusions is not a *further* commitment. The fusion is nothing over and above the cats that compose it. It just *is* them. They just *are* it. Take them together or take them separately, the cats are the same portion of Reality either way. (1991: 81)

(Imagine bargaining over two cats in a pet store. Can you buy the cats without buying their sum? Can you buy the sum but not the individual cats?[25]) To some extent, this was also the gist of Goodman's reply to the popular objection that a liberal sum postulate may have counterintuitive instances when the summands are scattered or otherwise ill-assorted, such as a cat and an umbrella, or a doughnut and its hole, or a sphere and its rotation.[26] (Or just think of Whitehead's view that there are no disconnected events.) We may well feel uneasy about treating unheard-of fusions as *bona fide* individuals, but this psychological fact has no bearing on the question of their ontological status. We may want to say that a sum cat+umbrella is not the same *sort* of thing as a cat, but that is no ground for doing away with the cat+umbrella altogether.[27] Besides, that policy would likely result in doing away with too many things. Everything is odd to a degree. James van Cleve put this point most clearly:

> Even if one came up with a formula that jibed with all ordinary judgments about what counts as a unit and what does not, what would that show? Not, I take it, that there exist in nature such objects (and such only) as answer the formula. The factors that guide our judgments of unity simply do not have that sort of ontological significance. (1986: 145)

One could consider adding further closure postulates, attesting for instance the existence of mereological differences and complements. In the presence of extensionality these notions can be defined as follows:

(3.18) $x-y =_{df} \iota z \forall w(Pwz \leftrightarrow (Pwx \wedge \neg Owy))$

(3.19) $\sim x =_{df} \iota z \forall w(Pwz \leftrightarrow \neg Owx)$.

In many versions, a closure theory also involves an axiom to the effect that the domain has an upper bound—that is, there is something (the universal individual) of which everything is part:

(3.20) $\exists z \forall x Pxz$.

Again, in the presence of extensionality such an object is unique and easily defined:

(3.21) $U =_{df} \iota z \forall x Pxz$.

This makes the algebraic structure of **CEM** even neater, since it guarantees that every two objects underlap and hence have a sum. On the other hand, few authors have gone so far as to postulate the existence of a "null individual" that is part of everything[28]:

(3.22) $\exists z \forall x Pzx$.

Without such a null individual (which one could hardly countenance except for algebraic reasons), the existence of a product is not always guaranteed, and (P.7) must be in conditional form. Likewise, differences and complements may not be defined—e.g., relative to the universe U.

3.5   Unrestricted Fusions

We may also add *infinitary* closure conditions. We may allow for sums of arbitrary non-empty sets of objects (and consequently also for products of arbitrary sets of overlapping objects: the product of all members of a class A is just the sum of all those things that are part of every member of A). It is not immediately obvious how this can be done if one wants to avoid commitment to classes and stick to an ordinary first-order theory. As a matter of fact, in some classical theories, such as those of Tarski (1929) and Leonard and Goodman (1940), the formulation of these conditions involves explicit reference to classes. We can avoid such reference by relying on axiom schemes that involve only predicates or open formulas. Specifically, we can say that for every satisfied property or condition ϕ there is an entity consisting of all those things that satisfy ϕ. Since an ordinary first-order language has a denumerable supply of open formulas, at most denumerably many classes (in any given domain) can be specified in this way. But this limitation is, of course, negligible if we are inclined to deny that classes exist except as *nomina*. We thus arrive at what has come to be known as Classical or General Mereology (**GM**), which is obtained from **M** by adding the axiom schema

(P.8) $\exists x \phi \rightarrow \exists z \forall y \, (Oyz \leftrightarrow \exists x \, (\phi \wedge Oyx))$ *(Fusion)*

(where ϕ is any formula of the language). The result of adding (P.8) to **EM** or **MM** yields correspondingly stronger theories. In fact, both **MM** and **EM** extend to the same strengthening of **GM**, which we call General Extensional Mereology (**GEM**).

It is clear that both **GM** and **GEM** are actually extensions of **CM** and **CEM**, since (P.6) and (P.7) follow from (P.8). Moreover, if the extensionality principle is satisfied then again at most one entity can satisfy the consequent of the postulate. Thus, in **GEM** we can define the operations of general sum (σ) and product (π):

(3.23) $\sigma x \phi =_{df} \iota z \forall y \, (Oyz \leftrightarrow \exists x \, (\phi \wedge Oyx))$

(3.24) $\pi x \phi =_{df} \sigma z \, \forall x(\phi \rightarrow Pzx).$

(P.8) then becomes

(P.8′) $\exists x \phi \rightarrow \exists z(z = \sigma x \phi),$

which implies

(3.25) $\exists x \phi \wedge \exists y \forall x(\phi \rightarrow Pyx) \rightarrow \exists z(z = \pi x \phi),$

and we have the following definitional identities (whenever the relevant existential presuppositions are satisfied):

(3.26) $x + y = \sigma z \, (Pzx \vee Pzy)$

(3.27) $x \times y = \sigma z \, (Pzx \wedge Pzy)$

(3.28) $x - y = \sigma z \, (Pzx \wedge \neg Ozy)$

(3.29) $\sim x = \sigma z\, (\neg Ozx)$

(3.30) $U = \sigma z\, (Pzz)$.

(It may be useful to compare these identities with the definitions of the corresponding set-theoretic notions, with abstraction in place of fusion.) This gives us the full strength of **GEM**, which is in fact known to have a rich algebraic structure: in 1935, Tarski showed that the parthood relation axiomatized by **GEM** has the same properties as the set-inclusion relation (more precisely, as the inclusion relation restricted to the set of all non-empty subsets of a given set, which is to say a complete Boolean algebra with the zero element removed).[29]

Various other equivalent formulations of **GEM** are also available, using different primitives or different sets of axioms. For instance, it is a theorem of every extensional mereology that parthood amounts to inclusion of overlappers:

(3.31) $Pxy \leftrightarrow \forall z\, (Ozx \rightarrow Ozy)$.

It follows that in an extensional mereology 'O' could be used as a primitive and 'P' defined accordingly. In fact, the theory defined by assuming (3.31) together with the Fusion Axiom (P.8) and the Antisymmetry Axiom (P.2) is provably equivalent to **GEM**, but more elegant. Another elegant axiomatization of **GEM** is obtained by taking only the Transitivity Axiom (P.3) and the Unique Fusion Axiom (P.8') (Tarski 1929). In the remainder of this work, however, we use our initial formulation for ease of comparison.

3.6 Atomism

The logical space of the mereological theories considered so far is schematically represented in figure 3.2.

We conclude this foray into mereology by briefly considering the question of atomism. Are there any mereological atoms—i.e., entities with no proper parts? And if there are atoms, is everything made up of atoms? These are deep and difficult questions, and in the chapters that follow we have various things to say about their relevance to the theory of spatial representation (both in the sense of a type 1 theory and in the sense of a type 2 theory). For the moment, let us simply point out that all options

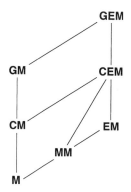

Figure 3.2
Hasse diagram of the basic mereological theories. (Inclusion goes uphill along the lines.)

are compatible with the mereological principles examined so far and can therefore be treated on independent grounds.

Concerning the first option, the assertion that there are no atoms amounts to the following:

(P.9) $\forall x \exists y PPyx.$ *(Atomlessness)*

Adding this axiom to a given mereological theory (generically labeled **X**) yields a corresponding *Atomless* version (labeled $\overline{\mathbf{A}}\mathbf{X}$). The second option clearly makes no sense in an atomless theory, for nothing can then be made up of atoms. However, in the absence of (P.9) one can consider adding an atomicity requirement to the effect that everything is ultimately composed of atoms:

(P.10) $\forall x \exists y (Pyx \wedge \neg \exists z PPzy).$ *(Atomicity)*

Call the result of adding this axiom the *Atomistic* version of a mereological theory **X** (**AX**). (P.9) and (P.10) are mutually inconsistent; however, it bears emphasis that any mereological theory considered so far is compatible with either position, and can therefore be made into an atomic or an atomless theory as the case may be.

Of course, any finite model of **M** (and *a fortiori* of its extensions) must be atomistic, since finitude together with the antisymmetry of parthood (P.2) jointly imply that decomposition into parts must eventually come to an end. Accordingly, every atomless model of **M** must have infinite cardinality. (A world containing such wonders as Borges's Aleph, where parthood is not antisymmetric, might by contrast be finite and yet atomless.) Moreover, atomistic mereologies admit of significant simplifications in the axioms. For instance, **AEM** can be simplified by replacing (P.5) and (P.10) with

(P.5') $\quad \neg Pxy \rightarrow \exists z(Pzx \wedge \neg\, Ozy \wedge \neg\exists w PPwz).$

Hence, it is apparent that an atomistic extensional mereology is hyperextensional in Goodman's sense: things built up from exactly the same atoms are identical.[30]

4 Connection Structures

We are now in a position to take a closer look at the idea of a topological extension of mereology—an extension that would take us beyond the prospects of a pure theory of parthood to deliver a theory of parts and *wholes*. The rationale for this move is especially clear if we consider **GEM** (General Extensional Mereology), and our informal arguments of chapter 2 were indeed given against the intuitive background of such a theory. The problem is this: In view of the fusion axiom (P.8), to every specifiable set of objects there corresponds a complete whole (their mereological fusion). Yet it is impossible in **GEM** to draw a distinction between "good" and "bad" wholes. There is no way to tell an integral whole from a scattered sum of disparate entities. Likewise, in **CEM** (Closure Extensional Mereology) the existence of a sum $x+y$ is conditional on the existence of an object z containing both x and y (P.6). But nothing says what further properties this object must have in order for the sum to exist. The sum of your left thumb and your left index finger might be more natural than the sum of your left thumb and your right index finger. But there is no way to express the difference in terms of mereological concepts. This was also the gist of our discussion of Whitehead's problem. Whitehead's early theories do not satisfy (P.6) or (P.8). For Whitehead, a necessary condition for two entities (events) to have a sum is that they be at least connected to each other. Yet we have seen that a purely mereological characterization of this connection relation falls short of capturing the relevant distinctions. Whitehead's proposed definition, which we can now formulate as in (4.1), works only if the bound variable 'z' is already assumed to range exclusively over self-connected entities:

(4.1) $Cxy =_{df} \exists z(Ozx \wedge Ozy \wedge \forall w(Pwz \rightarrow Owx \vee Owy))$.

(See again figure 2.2.)

It is here that topology comes into the picture: regardless of what specific mereological principles we assume, we must supplement them with an independent account of the relation of connection. To facilitate comparisons, let us expand our language by adding the predicate constant 'C' (connection) as a second primitive next to 'P'. The question of how mereology can actually be expanded to a richer part-whole theory may then be addressed by investigating how a P-based mereological theory of the sort outlined in chapter 3 can be made to interact with a C-based topological theory. This will be our task in the present chapter.

4.1 Entering Topology

As with mereology, we may ideally distinguish "lexical" from substantial postulates for 'C', regarding the former as embodying a set of minimal prerequisites that any theory purporting to explicate the meaning of the concept of connection must satisfy. Arguably, these must include the twofold requirement that 'C' be reflexive and symmetric[1]:

(C.1) Cxx *(Reflexivity)*

(C.2) $Cxy \to Cyx$.   *(Symmetry)*

By contrast, 'C' need not be transitive. France is connected to Italy and Italy to Slovenia, but France and Slovenia are disconnected.

   We call the basic theory defined by (C.1) and (C.2) *Ground Topology* (**T**), in analogy with the terminology of section 3.2. Of course this is an extremely weak theory, and a lot will have to be added before we can say that we have a topology. A model of **T** can be obtained simply by interpreting 'C' as the relation of mereological overlap (compare the **M**-theorems (3.8) and (3.9)), and what further principles should be added to (C.1) and (C.2) (so as to distinguish 'C' from 'O', for instance) is precisely the question a good topological theory is meant to answer. Since such further principles are controversial, we refrain from including them among the lexical axioms. For instance, should one assume connection to be extensional, so that things that are connected to the same entities are identical? Should one assume that any two connected things satisfy at least some form of Whitehead's principle (4.1)? Or consider the relation 'x is enclosed in y' defined by

(4.2) $Exy =_{df} \forall z\,(Czx \to Czy)$.

It follows from (C.1) and (C.2) that this relation is reflexive and transitive:

(4.3) $\mathrm{E}xx$

(4.4) $\mathrm{E}xy \wedge \mathrm{E}yz \rightarrow \mathrm{E}xz.$

And if C is extensional, than E is also antisymmetric (a partial ordering):

(4.5) $\mathrm{E}xy \wedge \mathrm{E}yx \rightarrow x=y.$

Should one assume this relation to satisfy any analogues of the axioms for 'P'? For each mereological predicate defined in (1)–(7) using 'P' we can now define a corresponding topological predicate using 'E'. Shall we assume any corresponding axioms?

As it turns out, it is difficult to answer these questions in an abstract setting.[2] Obviously, much depends on how exactly 'C' is interpreted, and obviously that depends in turn on how one thinks 'C' and 'P' should interact. Rather than pursuing these questions in isolation, then, let us proceed immediately to examining the main options for combining mereology and topology. There are essentially two strategies.[3] One can think of 'P' and 'C' as independent (though mutually related) predicates, or one can think of reducing one to the other. The first strategy—the one we favor—corresponds to the idea that mereology provides the foundation for richer, more sophisticated theories, of which topology is one (but not the only) important example. Tarski's (1929) work on the geometry of solids, for instance, proceeds by extending mereology (a version of **GEM**) with a primitive 'x is a sphere' to allow for a definition of solid geometric correlates of ordinary point-geometric notions. And Lejewski (1982) outlines a Leśniewskian theory of time (*Chronology*) in which a mereological basis is supplemented with a primitive relation 'x is wholly earlier than y' to account for the relation of temporal precedence. The theory of the relation 'x is located at y' that we consider in chapter 7 may be construed in the very same spirit. The second strategy corresponds to the idea that topology properly includes mereology. It stems from Whitehead's latest views (1929), which we briefly discussed in section 2.2. On this approach, just as mereology can be seen as a generalization of the fundamental theory of identity (insofar as parthood, overlapping, and even fusion subsume singular identity as a special case), topology could be viewed as a generalization of mereology, where the relation of connection takes over parthood and overlap as definable special cases. We have already put forward our misgivings about this way of proceeding in chapter 2, but a more rigorous account is nevertheless in order. We shall also consider a third possibility, which is a sort of vindication of mereology building on its ontological neutrality. On this view, topology may be viewed as a sub-theory of mereology, but as a *domain-specific* subtheory, topological no-tions being explained in terms of mereological relations among entities of a specified sort (e.g., extended regions).

4.2   Combining Mereology and Topology: The First Strategy

Let us begin with our favored strategy. The simplest account is straight-forward from the foregoing remarks: one could simply add the axioms (C.1) and (C.2) of Ground Topology (**T**) to the axioms (P.1)–(P.3) of Ground Mereology (**M**). However, this is of no interest unless one also adds some new principle bridging **M** and **T**. And the most obvious such principle is that parthood is a form of monotonicity: if one thing is part of another, whatever is connected to the first is connected to the latter—i.e., the first is topologically enclosed in the latter:

(C.3) $Pxy \rightarrow Exy$.   *(Monotonicity)*

This corresponds to the "obvious" direction of Whitehead's explication of parthood in terms of connection (see (2.2)), though we refrain from assuming also the other direction (enclosure as a sufficient condition for parthood). We call the resulting theory *Ground Mereotopology* (**MT**). To be sure, the axioms of **MT** are slightly redundant and could be simplified by replacing (C.2) and (C.3) with a single axiom:

(C.3′) $Pxy \rightarrow \forall z\,(Cxz \rightarrow Czy)$.

We keep (C.2) and (C.3) for greater perspicuity.

Note that (C.3) immediately implies that mereological overlap is a form of connection:

(4.6) $Oxy \rightarrow Cxy$.

However, the converse may fail, and we may take this failure as a defining condition for the relation of external connection—connection without sharing of parts:

(4.7) $\text{EC}xy =_{\text{df}} Cxy \wedge \neg Oxy.$

On the intended interpretation, this relation holds between things that barely "touch" or "abut" each other. It is symmetric, but neither transitive nor reflexive. Further mereotopological predicates can be defined as follows:

(4.8) $\text{IP}xy \;=_{\text{df}} Pxy \wedge \forall z(Czx \to Ozy)$ (Internal Part)

(4.9) $\text{TP}xy \;=_{\text{df}} Pxy \wedge \neg\text{IP}xy$ (Tangential Part)

(4.10) $\text{IO}xy \;=_{\text{df}} \exists z(\text{IP}zx \wedge \text{IP}zy)$ (Internal Overlap)

(4.11) $\text{TO}xy =_{\text{df}} Oxy \wedge \neg\text{IO}xy$ (Tangential Overlap)

(4.12) $\text{IU}xy \;=_{\text{df}} \exists z(\text{IP}xz \wedge \text{IP}yz)$ (Internal Underlap)

(4.13) $\text{TU}xy =_{\text{df}} Uxy \wedge \neg\text{IU}xy.$ (Tangential Underlap)

More generally, for each mereological predicate $\mathcal{R}$ we can define corresponding mereotopological predicates I$\mathcal{R}$ and T$\mathcal{R}$ by substituting 'IP' and 'TP' (respectively) for each occurrence of 'P' in the definiens. For instance:

(4.14) $\text{IPP}xy =_{\text{df}} \text{IP}xy \wedge \neg\text{IP}yx.$ (Internal Proper Part)

The patterns in figure 4.1 indicate how these predicates may represent a genuine addition to the purely mereological predicates introduced in (3.1)–(3.7).

Of course, the intended interpretation of 'C' is not yet captured by the axioms of **MT**; in this sense figure 4.1 is biased. Moreover, some of these patterns may be misleading. For instance, how exactly should the difference be represented between external connection and tangential overlap? The leftmost pattern gives an approximate representation of the former. But exactly what goes on at the dashed line, where these two externally connected entities meet? Imagine yourself proceeding along a path connecting the interiors of the two regions in the pattern. What happens as you leave x to enter y? Clearly, you do not pass from a last point of x to a first point of y; for if space is dense, then there would exist countless many intermediate points, contrary to the hypothesis that x and y are in touch. But if it is not possible for there to be two adjacent boundary points, shall we acknowledge the existence of only *one* such

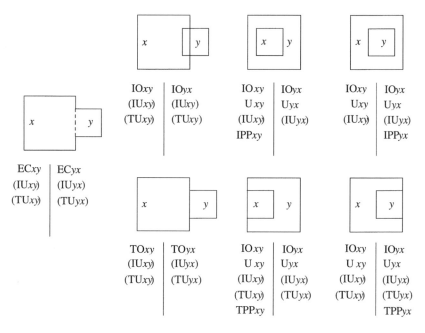

Figure 4.1
Basic connection patterns. As in figure 3.1, the clauses in parenthesis hold if and only if there is a z underlapped by x and y (internally or tangentially).

point? On which grounds could we then choose between a last point of x and a first point of y? Shall we rather deny the existence of boundary points altogether? On a closer look, the thin line in figure 4.1 conceals an array of deep puzzles. (Nor is this just a representation problem. Surely, the cat's tail is connected to the rest of its body—there is nothing separating them. And surely the two don't overlap—there are no common parts. So how are these two feline parts attached?) We come back to some of these matters later in this chapter and also—more fully—in chapter 5.

It also bears emphasis that predicates of the form $I\mathcal{R}$ and $T\mathcal{R}$, such as those defined in (4.7)–(4.14), are not the only novelties afforded by the new primitive. The enclosure relation 'E' defined in (4.2) is distinct from all mereological relations too. And using 'E', further relations can be introduced—as already mentioned—by taking the E-counterpart of the P-based relations defined in (3.1)–(3.7) or in (4.7)–(4.14). These relations would all be different from the relations of figure 4.1, as one can verify by interpreting 'Cxy' as 'the region of x and the region of y have at least one point in common'. If x and y are entities that may interpenetrate (two ghosts, two events), then they may stand in an E-based relation without standing in the corresponding P-based relation. (This is not the intended interpretation, but we cannot rule that out until we have a true theory of location.) One can also verify the independence of these relations by interpreting 'Cxy' as ordinary topological connection. If x is a point internal to a region and y is the region minus the point, then x is enclosed in y in spite of their being mereologically unrelated.

Finally, **MT** allows to define the predicate for self-connectedness that marks the difference between solid wholes (such as a table or a cup) and scattered wholes (such as a bikini or a broken glass). As was pointed out in section 2.2, the characterization of this predicate can be patterned after the ordinary topological definition, modulo a mereological rather than set-theoretical basis: a self-connected whole is one that cannot be split into two disconnected parts. More precisely, we shall say that x is self-connected if any two parts that make up the whole of x are connected to each other:

(4.15) $SCx =_{df} \forall y \forall z (\forall w (Owx \leftrightarrow Owy \lor Owz) \to Cyz)$.

4.3   Further Mereotopological Axioms and Principles

MT is hardly of any interest by itself. Though it allows us to define a number of mereotopological predicates, it does little to guarantee that these predicates receive an adequate interpretation. We may therefore consider the result of combining **T** with stronger mereological bases than **M**. Given any mereological theory **X**, we obtain a corresponding Mereotopology **XT** by adding the **T**-axioms (C.1) and (C.2) along with the bridge axiom (C.3). But even so, the resulting mereotopologies may be too weak unless further axioms for 'C' are added.

Consider for example **CMT**, the mereotopology induced by Closure Mereology (this amounts to taking **X**=**CM**).[4] In this theory, any two underlapping entities have a sum by (P.6). However, nothing guarantees that entities that are connected also underlap, not even when such entities are self-connected. Hence there is no guarantee that connected entities add up to a whole. Since connection is a rather natural condition for the existence of sums, one could therefore argue that a minimal mereotopological extension of **CM** should include, besides the basic axioms (C.1)–(C.3), also an axiom to the effect that connection implies underlap:

(C.4) $Cxy \to Uxy$.   (*Underlap*)

This would imply that if two self-connected entities are connected, they add up to a self-connected whole:

(4.16) $Cxy \wedge SCx \wedge SCy \to \exists z(SCz \wedge \forall w(Owz \leftrightarrow Owx \vee Owy))$.

   The same argument applies if we take **X**=**CEM**, the extensional strenghtening of **CM**. The only difference is that in the presence of extensionality (4.16) would amount to the more perspicuous

(4.16′) $Cxy \wedge SCx \wedge SCy \to SCx+y$,

and the definition of 'SC' would itself be simpler:

(4.15′) $SCx =_{df} \forall y \forall z\ (x=y+z \to Cyz)$.

   A more interesting case is obtained by taking **X** = **GEM**. The theory **GEMT** resulting by adding (C.1)–(C.3) was considered for the first time by Andrzej Grzegorczyk (1960) and has been advocated by Barry Smith (1993).[5] In this theory (C.4) is provable (trivially, since every pair of objects is underlapped by their sum, let alone by the universe). Moreover, using the full strength of the fusion axiom, in **GEM** it is possible to integrate the operators of sum, product, etc. with a variety of topological operators:

(4.17) $ix\ =_{df} \sigma z IPzx$ (interior)

(4.18) $ex\ =_{df} i(\sim x)$ (exterior)

(4.19) $cx\ =_{df} \sim(ex)$ (closure)

(4.20) $bx\ =_{df} \sim(ix+ex)$.   (boundary)

Like the mereological operators in (3.26)–(3.30), these operators are partially defined, unless we assume the existence of a null individual that is part of everything (see (3.22)). For instance, if x is a boundary, then it has no interior. And if x is the universal individual U, then it has no exterior. Even so, in **GEMT** these operators are all rather well-behaved. In particular, we can get closer to standard topological theories by supplementing (C.1)–(C.3) with the mereologized analogues of the standard Kuratowski (1922) axioms for topological closure[6]:

(C.5) $Px(cx)$ (*Inclusion*)

(C.6) $c(cx) = cx$ (*Idempotence*)

(C.7) $c(x+y) = cx+cy$.   (*Additivity*)

Equivalently, we could use the axioms for the interior operator:

(C.5′) $P(ix)x$ (*Inclusion*)

(C.6′) $i(ix) = ix$ (*Idempotence*)

(C.7′) $i(x \times y) = ix \times iy$.   (*Product*)

(These axioms are to be understood as holding whenever the operators c and i are defined for their arguments.) Each set of axioms is derivable from the other, so it makes no difference which one we take. Indeed, both (C.5) and (C.5′) are already implied by the corresponding definitions, (4.19) and (4.17), but we list them here for convenient reference.

We shall call the extension of **GEMT** obtained by adding these axioms *General Extensional Mereotopology with Closure conditions* (**GEMTC**). The strength of this theory is illustrated by the following consequences, which show in what sense the interpretation of 'C' in this theory is germane to that of standard set-theoretic topology: two things are (externally) connected if and only if they share (only) a boundary, i.e., if and only if the closure of one overlaps the other, or vice versa:

(4.21) $Cxy \leftrightarrow Oxy \lor Ox(cy) \lor O(cx)y$

(4.22) $ECxy \rightarrow Cxy \land \neg C(ix)(iy)$.

In view of such properties, **GEMTC** may be considered the archetype of a mereotopological theory based on the strategy under examination.

GEMTC also makes it possible to represent important mereotopological notions that escape the descriptive potential of weaker theories. For instance, reference to an object's interior allows one to refine the notion of wholeness introduced in (4.15). As it stands, in fact, the predicate 'SC' is still too general to capture the desired notion of an integral whole—a one-piece entity. On the one hand, one would need a stronger notion of connectedness, ruling out entities made up of pieces that are only "barely" connected, such as the sum of two spheres touching at a single point. In **GEMTC**, we may obtain such a stronger notion by the additional requirement that the entity's interior be self-connected too:

(4.23) $SSCx =_{df} SCx \land SCix$.

(The first conjunct will be superfluous on some plausible assumptions concerning boundaries, which we forgo for the time being.) On the other hand, one would need some means for expressing the distinction between self-connected entities in general (such as the bottom half of a ball) and self-connected *wholes* (the entire ball). After all, we are interested in connected wholes as a significant case of unity—namely topological unity. This means can be found in **GEMTC** by singling out those entities that are *maximally* strongly self-connected:

(4.24) $MSSCx =_{df} SSCx \land \forall y(SSCy \land Oyx \rightarrow Pyx)$

or, more generally, by singling out entities that are maximally strongly self-connected *relative to* some property or condition ϕ:

(4.25) $\phi\text{-}MSSCx =_{df} \phi x \land SSCx \land \forall y(\phi y \land SSCy \land Oyx \rightarrow Pyx)$.

For instance, suppose ϕ picks out the class of material objects; then (4.25) will single out the largest such objects among those that are all of a piece: the ball, say, as opposed to its bottom half, because the bottom half is part of the ball, but the entire ball is not part of any other connected material object. (Reference to ϕ may be crucial here; for insofar as everything is connected to its complement by (4.21), the only maximally self-connected entity in the sense of (4.24) might well end up being the entire universe.) These and similar facts make **GEMTC** a mereotopological theory particularly suited for the representation of spatial entities, although its topological structure is admittedly quite simple.

It may be observed that **GEMTC** is still too weak to guarantee certain intuitive facts. For instance, consider again Whitehead's principle (4.1). We have seen that this principle provides an inadequate characterization of 'C', and in fact the biconditional corresponding to (4.1),

(4.26) $Cxy \leftrightarrow \exists z(Ozx \land Ozy \land \forall w(Pwz \rightarrow Owx \lor Owy))$,

is false in **GEMTC** unless the quantified variable 'z' is assumed to range exclusively over self-connected entities. However, the predicate for self-connectedness is available in **GEMTC**, so one might want to consider a rectified version of Whitehead's principle:

(4.27) $Cxy \leftrightarrow \exists z(SCz \wedge Ozx \wedge Ozy \wedge \forall w(Pwz \rightarrow Owx \vee Owy))$.

(One could even consider using 'SC' as a primitive and define 'C' by the right-hand side of (4.27).) From right to left, this biconditional is indeed a theorem of **GEMTC**. If z overlaps both x and y as required, then $z \times x$ is connected to $z \times y$, and therefore x is connected to y by the Monotonicity Axiom (C.3). In the other direction, however, (4.27) may fail. For instance, if x has no self-connected parts, then there may be things to which x is connected (e.g., x itself) without there being any self-connected z doing the job required by Whitehead's principle. (An example of an entity with no self-connected parts is Cantor's bar: see figure 4.2.) Accordingly, this direction of (4.27) would have to be treated as an independent axiom:

(C.8) $Cxy \rightarrow \exists z(SCz \wedge Ozx \wedge Ozy \wedge \forall w(Pwz \rightarrow Owx \vee Owy))$.

We refrain from endorsing (C.8) at this point, but we may certainly conceive of applications for such an extended theory. Let us label any theory resulting from adding (C.8) to a given mereotopology **X** the *Whiteheadian extension* of **X** (**WX**). The thesis that everything has at least one self-connected part,

(4.28) $\exists z(SCz \wedge Pzx)$,

would then be a theorem of any such theory: it would follow by (C.2) taking x and y in (C.8) to be the same.

One final remark concerns the possibility of combining topological principles with either atomistic or atomless mereologies (see section 3.6).

Figure 4.2
Five steps in the creation of the Cantor bar. The first step is to remove the middle third of a self-connected bar. The next step is to remove the middle of each of the remaining bars. Repeating this over and over again creates a scattered object with no self-connected proper parts.

In particular, one may consider strengthening the assumption of mereological atomlessness to the requirement that everything has an *internal* proper part:

(C.9) $\forall x \exists y \text{IPP} y x.$ *(Boundarylessness)*

This would yield a mereotopology in which there exist no boundary elements such as lines, points, or surfaces. More precisely, where **X** is any mereotopological theory of the sort discussed so far, the *Boundary-free* variant $\overline{\mathbf{B}}\mathbf{X}$ of **X** is the extension of its atomless variant $\overline{\mathbf{A}}\mathbf{X}$ obtained by replacing the Atomlessness Axiom (P.9) with axiom (C.9). Obviously, in that case **X** cannot be an atomistic mereology, for (C.9) implies (P.9). And obviously $\overline{\mathbf{B}}\mathbf{X}$ (like $\overline{\mathbf{A}}\mathbf{X}$) can only have models with infinitely many elements.

More importantly, the chosen **X** must be weaker than **GEMT**, on pain of inconsistency. For in **GEMT** there is always a difference between an entity's interior and its closure (by P.5), and their difference (the entity's boundary) has no interior parts. In other words, the following conditionals are provable:

(4.29) $z \neq \text{U} \rightarrow \exists y(y = bz)$

(4.30) $y = bz \rightarrow \neg\exists x(\text{IP} xy).$

Whenever y and z satisfy the relevant antecedents, (C.9) fails.

This means that boundaryless theories are not easily accommodated within the present framework. As it turns out, most theories mentioned in this section are in fact committed to the existence of boundary elements.[7] At the same time, the admission of boundaries gives rise to some deep philosophical questions (such as those hinted at in our discussion of figure 4.1), so we shall have to come back to this issue when the overall picture is more complete (chapter 5).

4.4   The Second Strategy: Topology as a Basis for Mereology

Let us consider now the second way to bridge the gap between parts and wholes, corresponding to the idea that topology is a more general and fundamental framework subsuming mereology in its entirety. As was seen in section 2.2, the subsumption is obtained by defining parthood as topo-

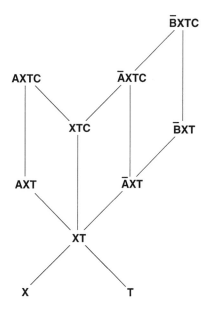

Figure 4.3
Hasse diagram of the mereotopological theories induced by a generic mereological theory
X. (Each theory has a Whiteheadian extension that is omitted from the diagram.) For
simplicity, *X* is assumed to be neutral with respect to atomism; also, in the text the closure
conditions defining *XTC* are given only for the case *X* = **GEM**.

logical enclosure, i.e., by assuming the basic mereotopological principle
(C.3) along with its converse:

(C.10) $Exy \rightarrow Pxy$.   *(Parthood)*

This gives us the basic characterization of this strategy: we shall call the
extension of **MT** obtained by adding (C.10) *Strong Mereotopology* (**SMT**).
This, in turn, can be simplified by observing that (P.1) and (P.3) are now
derivable from the other axioms. So in the end **SMT** is tantamount to the
theory defined by (C.1) and (C.2) along with the principle

(C.11) $\forall z(Czx \leftrightarrow Czy) \rightarrow x=y$,   *(Extensionality of C)*

a principle that corresponds to (P.2) with 'P' defined as in (2.2):

(4.31) $Pxy =_{df} \forall z(Czx \rightarrow Czy)$.

(This is actually the customary formulation of **SMT** found in the literature since Clarke 1981.[8])

This scarcity of primitives and axioms is admittedly a major attractive feature of **SMT**. However, it is equally possible to construct a mereotopology of the sort discussed in the previous sections using a single primitive. For instance, one can rely on the ternary relation 'x and y are connected parts of z'. Writing this as 'CPxy', one can easily define 'P' and 'C',

(4.32) $Pxy =_{df} \exists z CPxzy$

(4.33) $Cxy =_{df} \exists z CPxyz,$

and then go on to develop a mereotopological theory based on the relative independence of these two predicates.[9] (Note that (4.32) only presupposes connection to be reflexive—a perfectly unproblematic supposition—whereas (4.33) assumes all pairs of connected things to underlap each other—also a negligible supposition, regardless of how one feels about scattered entities.)

So the distinguishing feature of **SMT** is not its formal economy—the use of a single primitive. It is, rather, conceptual economy: the *notion* of parthood is fully subsumed under that of connection, that is, in the end, under the notion of wholeness. And the theory of parthood is, strictly speaking, a subtheory of the theory of wholeness. On this approach, the limits of mereology are overcome by turning the problem upside down: topology can deliver the full story about parts and wholes.

There is some cost too, though, and this was our main methodological objection to this strategy in chapter 2. This reduction may be conceptually convenient; but such a simplification—if acceptable—should be the result of a theory, not the starting point. For on what grounds can we decide at the very outset to identify parthood and enclosure? What interpretation of the connection predicate would warrant such an identification? In chapter 2 we saw that connection between two spatial entities cannot be explained in terms of connection between the corresponding spatial regions: we must provide a direct account. Consider then the following three possibilities:

(a) x and y are connected just in case they share a point.

(b) x and y are connected just in case x shares a point with the closure of

y or y shares a point with the closure of x (the standard interpretation of point-set topology).

(c) x and y are connected just in case their closures share a point.

The mereotopologies examined in the previous sections are all compatible with one or more of these readings. In particular, we have seen that option (b) is reflected in **GEMTC** by theorem (4.21). By contrast, in the presence of (C.10), each of these options yields implausible results.

Option (a) yields an implausible topology in which the boundary of a region is never connected to the region's interior (since they never share any points). For the same reason, on this account no object would be connected to its mereological complement, which is likewise implausible. (If they are not connected, *what* separates them?)

Option (b) yields an implausible mereology in which every boundary is part of its own complement (since anything connected to the former is connected to the latter). Among other things, this would imply that every object overlaps its complement, with a corresponding trivialization of the relation of external connection and, therefore, of mereotopology as a whole.

Option (c) yields an implausible mereotopology in which the interior of a region is always connected to its exterior (so that boundaries make no difference) and in which the closure of a region is always part of the region's interior.

Of course, one can avoid some of these results by banning all boundary elements from the domain. This amounts to extending **SMT** through the Boundarylessness Axiom (C.9). But it is just here that we have reservations. Why should our analysis of the notion of parthood force upon us a rebuttal of the boundary concept? Why should we assume (C.9) in order to ensure a coherent implementation of the Parthood Axiom (C.10)? And why should we reject all the readings in (a)–(c) above in order to avoid such outcomes? How is connection to be interpreted, if not according to one such reading?

These are our misgivings. Before we look closer into the exact nature of the connection relation (we are going to argue that (b) is indeed the preferred interpretation), let us complete our foray into this way of doing mereotopology by looking at some variants or extensions of **SMT**.

First of all, we may contemplate strong variants, not of **MT**, but of richer theories. Where **X** is any mereotopology of the sort discussed in the previous sections, the corresponding strong variant (labeled **SX**) is the extension of **X** obtained by adding the Parthood Axiom (C.10). For instance, **SWMT** is the strong variant of **WMT**, which in turn is obtained from **MT** by adding the Whiteheadian Axiom (C.8). (Curiously, Whitehead's (1929) own theory does not include (C.8), as an axiom or as a theorem. That theory is in fact (equivalent to) an extension of **SMT** obtained by adding various existential axioms such as the following:

(C.12) $\exists x \neg Cxy$

(C.13) $\exists x \exists y (Cxz \wedge Czy)$

(C.14) $\exists x \exists y (Pxz \wedge Pyz \wedge \neg Cxy)$.

We have no use of such principles in the following.)

Second, a theory based on (C.10) need not define the basic mereological operations in terms of 'O' (eventually construed in terms of 'C'). Rather, a common practice since Clarke 1981 is to make use of C-based definitions. This may mark a difference with regard to the '+' and '−' operators, for instance:

(4.34) $x +' y =_{df} \iota z \forall w (Cwz \leftrightarrow (Cwx \vee Cwy))$

(4.35) $x -' y =_{df} \iota z \forall w (Pwz \leftrightarrow (Pwx \wedge \neg Cwy))$.

More generally, define the C-based operator of general sum in this spirit:

(4.36) $\sigma' x \phi =_{df} \iota z \forall y (Cyz \leftrightarrow \exists x (\phi \wedge Cxy))$.

Then all mereological operators, and consequently all topological operators of the sort introduced in (4.17)–(4.20), are typically re-defined using (4.36) in place of (3.23). In the following, we use priming to indicate a theory exploiting such a way of proceeding (so that, for instance, **SGMT'** will be the strong variant of **GMT** modulo a systematic replacement of σ by σ'). A thorough comparison of these theories with their mereology-based counterparts would be very useful, and represents an open issue in the literature.

There are various other variants of the strong mereotopological strategy under examination. These include, for instance, the extension of the theory proposed by Aurnague and Vieu (1993a) (a version of **SB̄CMT'**), which has been motivated mostly by considerations concerning the semantics of natural language spatial prepositions,[10] and the theory proposed by Randell, Cohn, and Cui (1992b) (a variant of $\mathbf{S\overline{B}CEMT'}$) known as **RCC** (Region Connection Calculus), which has been motivated mostly by applications to spatial reasoning in Artificial Intelligence.[11] We need not look at these theories in further detail, though. Our purpose here was to give this strategy a rigorous formulation so as to facilitate comparisons with the strategy discussed in the previous section. But we have found reasons for preferring that strategy to the one under discussion, so we may pause here. Eventually we shall have more to say, and a uniform framework of reference, rather than a complete map of possible theories, is what we shall need.

4.5 Topology and Restricted Mereology

Are there other ways of construing a mereotopology besides the two main strategies examined above? Eschenbach and Heydrich (1995) have suggested that topology may be viewed as sort of restricted mereology. The two main strategies considered above have this in common—that they both take connection to be an extension of overlap, as reflected in theorem (4.6) (a consequence of (C.3), hence a theorem of all mereo-topologies considered so far). By contrast, on Eschenbach and Heydrich's proposal it is overlap that extends connection: for them connection is overlapping of *entities of a kind*, and the distinctive topological idea of external connection is made safe as long as the common part of two overlapping entities need not itself be a an entity *of that kind*. Using 'ϕ' to express the distinguishing property or condition of the kind in question, this amounts to assuming the following:

(4.37) $Cxy \leftrightarrow (Oxy \wedge \phi x \wedge \phi y)$

(4.38) $ECxy \leftrightarrow (Cxy \wedge \forall z(Pzx \wedge Pzy \rightarrow \neg \phi z))$.

Now, for Eschenbach and Heydrich, the kind in question is constituted by extended spatial regions, i.e., regions of space with a non-empty inte-rior. (In their system, the condition ϕ is represented by a simple predicate 'R', for 'is a region'.) Accordingly, on this interpretation (4.37) and (4.38) amount to saying that connection is region overlap, and that external connection is only possible between regions that share at most some boundary points. This was indeed Whitehead's (1929) suggested interpre-tation of 'C', corresponding to option (a) in section 4.4 above. And it is

the interpretation employed by Clarke (1981). So Eschenbach and Heydrich's suggestion is that a Whitehead-style mereotopology is not a strong
mereotopology (in our terminology). It does not explain mereology in
terms of connection. Rather, it explains connection in terms or the
mereology of regions:

> A topological concept which is even more basic than connection is region. . . . We
> claim that the distinction between regions and non-regions already allows for the
> establishment of the whole framework of [mereotopology] on the basis of [clas-
> sical mereology]. (Eschenbach and Heydrich 1995: 732)

This is legitimate because, of course, (4.37) can be used as a way of
defining 'C'. With 'C' thus defined, the ϕ-restrictions of all of (C.1)–(C.3),

(4.39) $\phi x \rightarrow Cxx$

(4.40) $\phi x \wedge \phi y \rightarrow (Cxy \rightarrow Cyx)$

(4.41) $\phi x \wedge \phi y \rightarrow (Pxy \rightarrow \forall z(\phi z \wedge Czx \rightarrow Czy))$,

would be provable as purely mereological theorems of **M**.[12] Hence the
Ground Topology theory **MT** would indeed be nothing but the subtheory
of **M** obtained by uniformly restricting the range of quantifiers by a
monadic condition ϕ—e.g., the property of being a region. Any extension
of **MT** would likewise involve, not a direct specification of 'C', but rather
a specification of the underlying concept of region. In particular, Eschenbach and Heydrich show how suitable axioms on 'R' make it possible to
specify Clarke's strong mereotopology (a version of **SBGMTC'**).

All this shows that, in a sense, a mereological theory needs very little
help in order to cope with certain basic topological notions and principles,
such as those discussed in this chapter. Formally it is only a matter of
restricted quantification. This is especially instructive insofar as the restricting condition ϕ in (4.37) need not be interpreted as expressing the
property of being a region—it can express a whole range of properties. In
particular, one could take it to express the property of being an extended
body, and this would dispose of all misgivings about a topology concerned
exclusively with spatial items rather than full-fledged spatial entities.

We welcome this sort of result. It bears witness to the richness of the
conceptual network linking mereology and topology. But we must emphasize: construing topology as restricted mereology is not a way of
solving Whitehead's problem without leaving the bounds of mereology.

Although such an account would not extend mereology by adding a topological primitive, it would still be necessary to add *some* primitive to express the relevant restricting condition φ. It can be a region predicate 'R' (suitably axiomatized), or it can be something else. But we must be able to say what subclass of entities in the domain of quantification can be connected, and what cannot. And this cannot be a purely mereological business.

As a way of combining mereology and topology into a unified theory, this strategy is therefore on a par with the first strategy considered above: both call for an extension of mereology. There is but one difference: that the present strategy regards topology as the business of a restricted kind of entities (the φers), whereas the first strategy has no restrictions whatsoever. To us this is a big difference indeed, but one that favors the first, unrestricted strategy. As a framework for investigating the structures of spatial representation, mereotopology cannot presuppose a thorough account of the φers. For the φers (whether these are just regions of space or spatially extended entities) are precisely the sort of thing whose part-whole structure we want to represent. The first strategy is the only one that serves this purpose adequately, and the one we follow in the remainder of this book.

-

5 Boundaries

We now take a closer look at one of the main theoretical options that were left open in our study of part-whole structures in the previous chapters, namely the nature of boundaries, and the many philosophical conundrums that come with it. The choice between a boundary-based and a boundary-free mereotopology can hardly be eluded, since the notion of external connection (connection without overlap) may change significantly depending on whether or not boundaries are included in the domain of reference. Boundaries seem needed in order to account for certain basic intuitions—to do justice to the ordinary concept of a surface, for instance. But they also give rise to deep puzzles that a boundary-free theory would escape altogether. Eventually we shall opt for the first attitude, treating boundaries as *bona fide* spatial entities along with ordinary objects. But we shall note that boundaries have a peculiar relation to space: they are located in space, yet do not take up any space (just as temporal boundaries do not take up any time). This turns out to be a deep distinction.

5.1 Boundary Puzzles

The concept of a boundary arises as soon as we think of an object demarcated from its environment. There is a boundary demarcating France from Germany; there is a boundary (a surface) demarcating the interior of the table from its exterior. Often, things have boundaries that are physically salient and perceivable, such as skins and crusts. Other things seem to have fuzzier boundaries. (Where does the hinterland begin? What exactly are the boundaries of that cloud?) But whether sharp or fuzzy, for every object there appears to be a boundary demarcating it from the rest of the world. Events too have boundaries, at least temporal boundaries: Susan's swim through the Channel began with her dive and ended with her last stroke; our lives are bounded by our births and our deaths.

This picture is so natural and pervasive that it is hard to deny boundaries a central place in our conceptual scheme. Not only do they enter the content of our perceptions. (We seem to *see* boundaries in a number of places). They belong to the palette of basic ontological tools that we commonly use to describe the spatiotemporal world. Yet some deep and difficult questions arise as soon as the duty of boundaries is taken seriously into account. A boundary demarcates two entities, or two parts of the same entity, which are said to be in contact with each other. How is this contact to be explained? Consider the tail of a cat and the rest of its body. Shall we say, following Brentano (1976), that there are *two* boundaries (one belonging to the tail and one belonging to the rest of the cat's body) that share exactly the same spatial location? Or shall we rather follow Bolzano's (1851) doctrine (a "monstrous doctrine," according to Brentano 1976: 146) and maintain that there is only *one* boundary—a boundary that belongs to one of these two entities (the tail or the body) but not to the other? Intuition has no straight answer to such questions. Yet there is no point in denying that these questions define important choices to be made by any theory of boundaries—or any boundary-based theory of spatial representation.[1]

A first methodological remark is in order. We might put it as follows: our primary concern is with the ordinary conception of middle-size reality. But this does not mean that we are interested in taking a stand on the issue of whether middle-size objects have true boundaries, if that is understood on the assumption that physical (microscopic) reality is all the reality there is. We can assume that it is an obvious and uncontroversial fact that ordinary physical objects (i.e., objects intepreted physically as aggregates of molecules) are not strictly speaking dense and do not have boundaries of any kind (at least, not boundaries of the smooth, continuous sort countenanced by our unreflected view of the world). If the solid bodies of common sense are replaced by intricate systems of subatomic particles, speaking of a body's continuous boundary is like speaking of the "flat top" of a fakir's bed of nails, to borrow Peter Simons' phrase (1991c: 91). Boundaries become merely imaginary entities enveloping smudgy bunches of hadrons and leptons, and their exact shape and properties involve the same degree of arbitrariness as those of any mathematical graph smoothed out of scattered and inexact data. We may take this to be a true story. But we are not interested in the problem of how middle-size boundaries "emerge" out of microscopic discontinuities, just as we are

not interested in how the top of a fakir's bed can—in some circumstances—be described as flat. Moreover, the physical framework drops
out as largely irrelevant in the folk justifications we provide for the
content of our perception and action. Mary decided to leave the highway
because she thought that the road was bumpy, and she thought so because
she saw it (and felt it) as uneven. This is the kind of description we use in
our everyday justifications, and it is also the description of the world we
need to rely on in order to explain the behavior of a cat or to teach a
machine—say, a robot with suitable sensory modules—to move around
and interact autonomously with the environment. As J. J. Gibson put it:

> There is physical structure on the scale of millimicrons at one extreme and on the
> scale of light years at another. But surely the appropriate scale for animals is the
> intermediate one of millimeters to kilometers, and it is appropriate because the
> world and the animal are then comparable. (1966: 21f)

It is this intermediate "world" that we are concerned with. We are uncommitted with regard to its status (whether it should be reduced to, supervene on, or added on top of an underlying physical reality). But we are
interested in describing its spatial structure, and in this respect objects are
conveniently conceptualized as finite chunks of dense matter with closed,
continuous boundaries.[2]

We take a similar stance with respect to the domain of events and
event-like temporal entities. Surely, even if we accept such entities as part
of our basic ontological inventory, one might argue that no *instantaneous*
events really exist, hence no boundaries are ever to be found in the realm
of happenings except in some loose sense. Thus, for instance, Antony
Galton remarks that, in a scientific spirit, a body's being at rest amounts
to the fact that the vector sum of the motions of the trillions of restless
atoms of which the body is composed, averaged over time, equals zero:

> The averaging over time makes it impossible to pinpoint, even in principle, an
> instant at which this condition ceases to obtain. Thus it might be claimed that no
> real meaning can be attached to the assertion that the body's beginning to move
> is an instantaneous event. (1994: 4)

As Galton points out, however, in the context of a qualitative theory this
objection is beside the point. This is not to insist on the somewhat questionable epistemological claim that we do perceive or have experience of
events that obtain for only an instant. Rather, the objection is beside the

point simply because the notion of two states or events abutting each other is part of our conceptual scheme, just as the notion of two objects touching each other. And whether or not one countenances instants, if one allows for abutting states, that which happens at their juncture is perfectly describable as the double-barrelled instantaneous event of one state's ending and the other state's beginning to hold.

We do not mean to insist too much on the naive-physical flavor of all this and, as we said, we do not intend to address the issue of the cognitive and ontological status of the middle-size world we attribute to common sense. Simply, much of what one can say about boundaries would be meaningless except on the grounds of such caveats. And even if talk of boundaries and contact is deemed unsuited to the ontology of the physical sciences, one still needs it when it comes to the dense entities carved out by ordinary discourse and to the spatial regions that these occupy.

5.2 Constraints on Theories

Boundaries presuppose density, then. But density is also a source of puzzles for boundaries, as the Brentano-Bolzano dilemma indicates. (What, if anything, divides the tail from the rest of the cat?) More precisely, we may locate the difficulty in the seeming joint incompatibility of the following three intuitions about space and spatial entities:

(a) density of space

(b) existence of boundaries

(c) possibility of contact between discrete (non-overlapping) entities.

Any attempt to subject the notion of boundary to systematic investigation must explain how the compatibility can be restored, or else give up one or more of (a)–(c).

We have just seen that giving up density is hardly a solution if we are interested in ordinary objects—so (a) should better stay. And we take it that denying (c), the possibility of contact between discrete entities, is likewise too radical a solution. To be sure, natural language does not distinguish between true topological connection and mere physical closeness. In general, the surfaces of two distinct physical bodies (such as John and Mary as they shake hands) cannot be topologically connected, though the bodies may of course be so close to each other that they appear to be in contact to the naked eye. This, however, leaves the question open in the case of two adjacent parts (the tail and the rest) or in those cases where the two candidates for contact are not physical objects. At the very least we want to say that every object must be in contact with its complement. But even this is enough to cause problems, as with Peirce's puzzle (1893: 98): Which color is the line of demarcation between a black spot and a white background? Similarly, in the temporal realm there arises Aristotle's riddle: At the instant when an object stops moving, is it in motion, or is it at rest? (*Physics* VI, 234a ff.) In both cases, it seems that any way of privileging one answer over the other, hence any way of deciding where the boundary belongs, leaving one entity closed and the other open, would involve a contravention of the principle of sufficient reason. So the burden is really on (b), and the acceptance of boundaries seems the only questionable intuition.

We take this to amount to the following alternatives.

One could say that boundaries are part of the ontology of common sense, but only of the *prima facie* ontology. Once we subject them to critical analysis we realize that they are merely a *façon de parler*. And as soon as we dispense with them, we also get rid of all the problems they carry along. One possibility was suggested by H. H. Price:

'Surface', it is true, is a substantive in grammar; but it is not the name of a particular existent, but of an attribute. (1932: 106)

This attributive shift, which de-reifies boundaries, is not an unpopular strategy (we have seen it in relation to the hole problem in section 2.3), but it entails some radical revisions in the semantics of natural language. The question then arises of how to account for the meaning of sentences quantifying over or referring to such fictitious entities. We call theories that follow this approach *boundary-free* theories, regardless of their specific revisionary content.

On the other hand, a truly boundary-based theory will have to take the above problems at face value and provide some direct account of the fundamental relation of contact (and, more generally, topological connection) in a dense domain: an account of (c) above in the presence of (a). This is the strategy we favor, and we shall eventually consider some ways of pursuing it. First, however, we want to assess the former option.

5.3 Boundaries as *façons de parler*

The view that boundaries are a mere *façon de parler* can be traced back to Whitehead's theory of "extensive abstraction."[3] We may see two aspects of the theory, both of which are significant for our purposes. On the one hand, the theory purports to provide an account of the connection relation. On the other hand, Whitehead's theory also aims to provide an indirect account of such notions as point, line, and surface in such a way as to do justice to their customary geometrical properties. That is, boundary elements are not included among the primary entities, but they are nonetheless retrieved as higher-order entities (which explains in what sense they are *façons de parler*). This is achieved through Whitehead's method of extensive abstraction, which effectively treats boundaries as classes of nested objects that "converge" (though not to an object proper).

Concerning this second aspect, there is indeed some appeal in the thought that the notion of a boundary involves some sort of abstraction.[4] Admittedly, 'abstraction' covers a mixed bag. One can find, in a list of abstract entities, numbers along with beliefs and geometrical figures, entities whose common feature is the absence of a concrete dimension. However, even though it is clear that boundaries might be taken as immaterial, it is equally clear that boundaries do not immediately satisfy some of the more obvious criteria for abstractness (as listed, e.g., in Künne 1983) such as unperceivability or atemporality. This is a source of concern for abstractive reductions of boundaries, especially insofar as these theories are supposed to solve the epistemological problem of how we perceive objects that do not have material parts. In particular, it is hard to find satisfaction in the idea that the abstractness of boundaries is to be explained in terms of the abstractness of set-theoretic constructions. One can see and paint the surface of a table, and one can even see and paint an infinite series of ever thinner layers of table-parts. But one cannot paint the *set* of these parts (unless of course this is simply another way of saying that the *parts* are painted). Indeed, Theodor De Laguna, one of the very first sponsors of Whitehead's method of extensive abstraction, remarked that the identification of points and other boundaries with classes of solids is "open to serious misinterpretation":

Instead of the point, which we do not perceive, we are given a class of solids such as we do perceive. . . . But . . . although we perceive solids, we perceive no abstractive sets of solids; and there is no indirect empirical assurance that such sets exist—only suggestive evidence that entitles us to assume they exist. In accepting the abstractive set, we are as veritably going beyond experience as in accepting the solid of zero-length. (1922c: 460)

If one wants to go boundary-free, then, one must independently make sense of all boundary-talk, regardless of whether one can then reconstruct boundaries as second or perhaps even higher-order entities. This, by the way, applies not only to Whitehead's account, but also to other techniques for recovering the notion of unextended boundaries in terms of extended, voluminous entities, whatever their ultimate underlying motivations.[5] (For instance, a boundary is completely individuated by the things that it separates, so one could think of construing the boundary of any given entity x as the set $\{x, \sim x\}$.[6])

Observe that we are thinking here of accounts of the geometric aspect of boundaries. An "operationalist" account of the sort advocated by Adams (1996), where the abstractive process by which boundary elements are derived from concrete observables is explained in terms of "operational tests", arguably eschews this line of objection. But such an account is best regarded as a parallel story, one that offers an explanation of empirical knowledge concerning boundaries while remaining ultimately neutral with regard to the their structural role in spatial representation.

5.4   The Incongruities of Boundarylessness

Let us then focus on the first task that a boundary-free theory must accomplish—an account of the connection relation. For the sake of precision, we consider Clarke's (1985) formulation, which is based on the theory of Strong Mereotopology (**SMT**) of chapter 4. We have seen that this sort of theory takes parthood to be definable in terms of connection. However, not much will depend on this feature of **SMT**, so we can leave this issue on a side here and focus on the question, What account of contact do we get in a boundary-free theory? In the absence of boundaries the relation of tangential overlap (TO) is obviously empty; but what is the intended model for the relation of external connection (EC)? What is it that marks the difference between connected and disconnected entities?

It is worth noting that Clarke's own theory is stronger than **SMT**. It includes the axioms for topological closure (though in a different format than our (C.5)–(C.7)) as well as the C-based version of the unrestricted Fusion Axiom (P.8) (see (4.36)), making it possible to introduce corresponding operators of complement, interior, exterior, and closure. Accordingly, nothing in principle would prevent one from also introducing an explicit boundary operator, in perfect agreement with standard topological treatments.[7] We have seen one definition in chapter 4:

(5.1a) $bx =_{df} \sim (ix + ex)$.

That is, the boundary of x is the complement of the sum of the interior and the exterior of x. But we have several other options—for instance,

(5.1b) $bx =_{df} cx \times c(\sim x)$

(5.1c) $bx =_{df} \sim (ix) - i(\sim x)$

(5.1d) $bx =_{df} \sim (ex) - e(\sim x)$.

Such definitions are not unproblematic if the relevant mereotopological operators (especially the complement operator $\sim$) are based on 'C' rather than 'O', and we shall come back to this shortly. First, however, let us make clear that anything along the lines of (5.1a)–(5.1d) is in fact ruled out explicitly by Clarke by means of the Boundarylessness Axiom:

(C.9) $\forall x \exists y IPPyx$.

Introducing this axiom makes the above definitions ineffective, i.e., bx is always undefined. For instance, the complement of the sum of the interior and the exterior of any given x has no interior parts, and is therefore not included in the domain of quantification. Hence (5.1a) is never defined.[8]

Formally, then, Clarke's theory is a version of the Strong variant of Boundary-free General Mereotopology with Closure Conditions (and C-based fusions), i.e., a version of **SBGMTC′**. This being the general picture, there are two main objections to it, each of which we find conclusive enough to suggest abandoning this route.

The first objection is that the banishment of boundaries collides with the fact that the theory allows for the standard topological distinction between "closed" entities (entities that contain their boundaries as parts)

and "open" entities (entities that contain no boundary parts, i.e., are equal with their own interiors). The usual definitions apply:

(5.2) $\text{Op}x =_{df} \ x = ix$ (open)

(5.3) $\text{Cl}x =_{df} \ x = cx.$ (closed)

More generally, and without appeal to the general fusion axiom, the theory has room for the distinction between interior and tangential parts. The distinction is actually needed to state the Boundarylessness Axiom (C.9), and is reflected in a theorem to the effect that interior parts (and hence open entities) are never externally connected to anything:

(5.4) $\text{IP}xy \rightarrow \neg \text{EC}xz.$

Now there is nothing wrong with this fact in and of itself, except that it becomes very difficult to make good sense of these distinctions. As Simons put it:

> What we are being asked to believe is that there are two kinds of individuals, "soft" (open) ones, which touch nothing, and partly or wholly "hard" ones, which touch something. Yet we are not allowed to believe that there are any individuals which make up the difference. We can discriminate individuals which differ by as little as a point, but are unable to discriminate the point. (1987: 98)

In other words, an open individual is always a proper part of its own closure, but there is no mereological difference between the two. This means that if you cut something open and take its interior, you get a proper part of the initial thing; yet you leave out *nothing*. In short, the Weak Supplementation principle of Minimal Mereology,

(P.4) $\text{PP}xy \rightarrow \exists z \, (\text{P}zy \wedge \neg \text{O}zx),$

fails whenever x is open and $y = c(x)$.

It is indeed hard to find satisfaction in this picture. Formally, what it amounts to is that the preclusion of boundaries from the domain of interpretation gives rise to a non-standard *mereology*. This is not surprising, and one may argue that it is the price one has to pay on this account: if some elements are removed from the domain, the supplementation principle cannot generally hold. (One can make good sense of this in the Eschenbach-Heydrich account discussed in section 4.5.) But the problem is that the resulting account of contact is heavily affected by this. For if we cannot pinpoint anything that explains the difference between things that touch and things that don't, then we can hardly say we have a notion of what it means for two things or regions to be in touch. It is this that we find unacceptable.

This ties in with the second objection. We take it that a theory of connection must be capable of explaining, not only the difference between external connection and connection with overlap, but also the difference between what may be called 'weak' and 'strong' external connection—between contiguity and continuity. Think of two spheres just touching each other (for simplicity, think of these as two spherical regions, connected to each other in the precise sense of sharing exactly one point). The way they are connected is qualitatively different from the way two halves of a single sphere are connected. In mereotopological terms, the difference is that in the first case (contiguity), one can go from any one part of one sphere to any part of the other without ever *going through the exterior* of the whole: this is what makes them connected. In the second case (continuity), one can go from any part of one half to any part of the other half without ever *leaving the interior*.

Now, this difference can easily be expressed within our favored boundary-based mereotopological apparatus (**GEMTC**). First of all, we have a distinction between the connectedness of a whole ('SC') and the connectedness of its interior, or strong connectedness ('SSC'). (See definitions (4.15) and (4.23), respectively.) Using this, we can then define a relation of strong connection ('CS') as holding between two items when their sum is strongly self-connected—more generally, when some parts of theirs add up to a strongly self-connected sum:

(5.5) $\text{CS}xy =_{\text{df}} \exists w \exists z (\text{P}wx \wedge \text{P}zy \wedge \text{SSC}(w+z))$.

This relation would have the desired characteristics: it would hold between two halves of a single sphere, but not between two spheres that are in contact—the whole constituted by the two spheres would be self-connected, but not strongly so. We can therefore define continuity:

(5.6) $\text{CN}xy =_{\text{df}} \text{EC}xy \wedge \text{CS}xy$.

Thus, CN would imply EC, but not vice versa, and we could say with Aristotle that "the continuous is a species of the contiguous." (*Metaphysics*, K 1069a)

All of this would be quite natural in a boundary-based topological framework such as **GEMTC**, for in this theory external connection is sharing of a boundary: two things are externally connected when they do not overlap but the closure of one overlaps the other (or vice versa) (4.21). Hence, external connection can only hold between a closed entity and a (semi-) open one (4.22). (This of course is the source of Pierce's puzzle, to which we shall have to come back.) But what about the Whitehead-Clarke approach? Here both halves must be closed, for otherwise they would not connect at all by (5.4). But then, if the two halves of a single sphere are both closed, we have no means of distinguishing their way of being connected from the way the two spheres are connected. Hence the desired distinction between continuity and contiguity is lost and we have a case of explanatory deficiency in the theory.

It bears emphasis that this is not simply a matter of changing the relevant definitions. Indeed, in Clarke's theory some definitions must be emended, for the Parthood Axiom (C.10) together with the Boundarylessness Axiom (C.9) have the implausible consequence that the universe can never be self-connected:

(5.7) $\neg SCU$.

This is so because the universe can always be split into two halves, say x and its complement, the interiors of which are disconnected and yet add up to the whole. The interiors and the universe are connected exactly to the same things, for there literally is *nothing* to make up for the difference:

(5.8) $\forall x(ix+ex = U)$.

We have here another manifestation of the failing of the Weak Supplementation Axiom (P.4). For himself, Clarke (1985: 69) eventually bypasses the problem by employing the following weaker definition of 'SC':

(5.9) $SCx =_{df} \forall y \forall z \, (y+z = x \to (C(cy)z \lor Cy(cz)))$.

This notion can be further weakened by requiring that the closures of both halves be connected (Asher and Vieu 1995):

(5.9′) $SCx =_{df} \forall y \forall z \, (y+z = x \to C(cy)(cz))$.

Such amendments would guarantee that 'SC' receive its intended interpretation. Yet the amendments would have no effect on the distinction between weak and strong contact. We would still need to account for the distinction somehow. And we would still need to account for of the "old" notion of self-connectedness expressed by (4.15) or (4.15′), which continues to make perfectly good sense no matter how we go about with the official nomenclature.

5.5 A Variant Treatment

Another feature of the Whitehead-Clarke theory that we find unpalatable is that nothing is connected to its own complement. This is obvious if we define the complement operator '~' in terms of 'C', as the theory actually has it:

(5.10) $\sim x =_{df} \sigma' z \neg Czx$.

(See (4.36) for the definition of σ', the C-based operator of general sum.) But we would have a similar result even if we employed the standard definition of '~' in terms of 'O':

(5.10′) $\sim x =_{df} \sigma z \neg Ozx$.

In this case, the theory would still imply that an entity is connected with its complement only if it is closed, giving rise to the embarrassing question of *what* separates an open entity from its complement (and to the question of how two distinct entities—an open entity and its closure—can have one and the same complement).

A variant of the theory in which this difficulty is overcome is the calculus **RCC** of Tony Cohn and others, which was mentioned briefly at the end of section 4.4.[9] This is essentially a version of the Strong Boundary-free Closure Mereotopology (**SBCMT**) motivated by a weaker interpretation of 'Cxy' as meaning 'the topological closures of x and y share a point'. In a sense, this feature makes **RCC** very akin to Asher and Vieu's account, as both mark a similar departure from **SBCMT**: while (5.9′) changes the basic notion of self-connectedness by replacing the clause 'Cyz' by '$C(cy)(cz)$', **RCC** builds that change directly into the intended interpretation of the C-relation. However, this shift of interpretation carries several significant consequences, making the departure rather substantial. In particular, the difference is reflected

formally in the employment of the following weaker notion of complementation:

$(5.10'')$ $\sim x =_{\mathrm{df}} \iota y \forall z ((Czy \leftrightarrow \neg \mathrm{IPP}zx) \wedge (Ozy \leftrightarrow \neg Pzx)).$

This says that the complement is (i) disconnected from every interior
proper part and (ii) discrete from every part of the object. It follows,
therefore, that every non-universal entity is connected (in effect, externally connected) with its own complement:

(5.11) $x \neq U \to ECx(\sim x).$

But this seemingly minor change has some major consequences. For, on
the one hand, it avoids the first above-mentioned objection to Clarke's
original formulation: the weak supplementation principle (P.4) is now a
theorem. On the other hand, $(5.10'')$ also implies a dissolution of the
distinction between open and closed entities. This follows from the fact
that O becomes an extensional relation, i.e., the following **EM**-theorem is
now provable:

(5.12) $x = y \leftrightarrow \forall z (Ozx \leftrightarrow Ozy).$

Thus, it is not possible for two items to share exactly the same parts
without having the same relationships in terms of connection, and in the
presence of the Boundarylessness Axiom (C.9) this deprives the
open/closed distinction of any foundation. The motivations adduced in
support of this change are instructive, as they reflect precisely the reason
why one would want to go boundary-free in the first place:

> From the naive point of view, the distinction between open, semi-open and closed
> regions is not drawn. . . . Moreover, . . . if we map bodies to closed regions (as the
> spaces they occupy), then their complements become open, which is a less agree-
> able result. An elegant solution to this problem becomes less likely once we allow
> bodies to break up into parts, since one is left with the result that one part will
> map to a closed region, and the other to a semi-open region! (Randell, Cui, and
> Cohn 1992a: 394–395).[10]

We don't think these improvements of the theory over Clarke's formulation are sufficient to give credibility to the idea of a boundary-free
mereotopology. For one thing, we are still left with the problem of accounting for the intuitive opposition between continuity and contiguity.
But second, and more important, we are left with the problem of making

sense of certain definable distinctions. For example, although the authors do not do so, nothing prevents us from introducing the interior and closure operators. In a sense, these would be redundant, in that (P.4) and (5.12) jointly imply the following identities:

(5.13) $x = \mathrm{i}x = \mathrm{c}x.$

But since everything is connected with its own complement, this amounts to saying that the sum of the interior parts of anything x is in contact with the complement of x. And this runs against any pre-theoretical intuition concerning 'sum' and 'interior part'. Likewise, since (the closure of) any region includes the sum of the region's tangential parts, (5.13) says that by putting together the interior parts of any region we eventually get all of its tangential parts—and this, again, goes against the purported intuitive content of these notions. One way or the other, a conceptual discrepancy threatens here.

Randell, Cui, and Cohn note this outcome in a later paper and conclude that the theory must therefore be subjected to "an important metatheoretic restriction"—namely that infinite sums cannot be allowed (1992b: 172). If we don't allow for infinite sums, we cannot define $\mathrm{i}x$ or $\mathrm{c}x$, hence the troubles are over. But why should the sum principle be called into question here? Why should mereology suffer from a topological disease? Besides, as was seen in chapter 3, there is some ambiguity in the notion of restricting sum formation. If we agree with the idea that the sum is nothing over and above the things that compose it (i.e., the sum is those things taken collectively), then the problem is not solved by formally ruling out sums of interior parts. For the interior parts are *all* there already. It is hard to see any reason why *these* fusions should be proscribed—except for their responsibility in unveiling the limits of the Boundarylessness Axiom (C.9).

5.6 Accepting Boundaries

Now let us consider the second possibility mentioned at the end of section 5.2, namely that boundaries must be included in the inventory of spatial entities in spite of their seeming incompatibility with either density or the possibility of contact. To this end, let us recall that in our favored mereotopological apparatus (**GEMTC**) boundaries can be defined explicitly using any of the definitions in (5.1a)–(5.1d). We also have some other options, including some that may be more illuminating. For instance, we may define the boundary operator 'b' in terms of a more general relational boundary predicate 'B':

(5.1e) $bx =_{df} \sigma z\ Bzx.$

Intuitively 'B' is to be interpreted as holding whenever the first argument *bounds* the second and is thus *a* boundary of the second, though not necessarily *the* boundary of it. Since anything that bounds an entity must be part either of the entity or of its complement, we may therefore characterize 'B' through the auxiliary notion of a boundary part ('BP'):

(5.14) $BPxy =_{df} \forall z(Pzx \rightarrow TPzy)$

(5.15) $Bxy =_{df} BPxy \vee BPx(\sim y).$

Alternatively, we may characterize 'B' through the relation of straddling ('S'). Say that x straddles y if every neighbourhood of x (i.e., everything including x as an interior part) over-crosses y (see (3.4)). Then a boundary of an item y is any item x all parts of which straddle y[11]:

(5.16) $Sxy =_{df} \forall z(IPxz \rightarrow OXzy)$

(5.15′) $Bxy =_{df} \forall z(Pzx \rightarrow Szy).$

In **GEMTC** the two characterizations of 'B' are logically equivalent, so (5.1e) does not depend on the choice between (5.15) and (5.15′). More generally, it can be shown that on both readings (5.1e) is equivalent to any of (5.1a)–(5.1d), provided only we assume 'b' to satisfy the basic properties of the standard boundary operator:

(5.17) $bx = b(\sim x)$

(5.18) $b(bx) = bx$

(5.19) $b(x \times y) + b(x+y) = bx + by.$

(Jointly, these three properties are tantamount to the properties of 'c' expressed by the closure axioms (C.5)–(C.7), and to the properties of 'i' expressed by (C.5′)–(C.7′).[12]) To close the circle, we could also take either 'b' or 'B' as a primitive and introduce 'C' by definition, as per the following equivalences:

(5.20) $Cxy \leftrightarrow Oxy \lor Ox(by) \lor O(bx)y$

(5.21) $Cxy \leftrightarrow \exists z ((Pzx \land Pzy) \lor (Pzx \land Bzy) \lor (Bzx \land Pzy)).$

In both cases, a corresponding characterization of external connection follows immediately:

(5.22) $ECxy \leftrightarrow \neg Oxy \land (Ox(by) \lor O(bx)y)$

(5.23) $ECxy \leftrightarrow \exists z (\neg(Pzx \land Pzy) \land ((Pzx \land Bzy) \lor (Bzx \land Pzy))).$

Now, it is here that the puzzles discussed at the beginning arise. This way of chartacterizing connection requires a full endorsement of the Bolzanian opposition between open and closed entities: when two entities are externally connected, they share a boundary; but this sharing is uneven. The boundary only belongs to one object, and bounds the other *from the outside*. (More precisely, the sharing may be even, in that each of two externally connected objects may include half of the common boundary. But no part of the boundary can be a boundary part of both objects: where the boundary belongs to one, the other is bound from the outside.) So the question must now be faced: What grounds are there to classify one object as closed and the other as open (in the relevant contact area)? More generaly, what grounds are there to distinguish between closed and open entities in the first place?

Let us first of all clear the ground from a possible misapprehension. The open/closed distinction is problematic, but it is not in and of itself a "monstrous" distinction (as Brentano put it). In some cases we find it quite reasonable: ordinary material objects are naturally the owners of their boundaries (their surfaces, in effect), and there is nothing counter-intuitive in the thought that the environments in which they are embedded are open.[13] Or consider events: we can think of Vendler's (1957) classification of event types as introducing a distinction between open processes (activities such as Mary's swimming) and closed processes (accomplishments such as Mary's crossing of the English Channel), depending on whether the relevant boundary, or achievement (Mary's reaching of the French coast), is included in the event to which reference is made.[14]

Let us then focus on the main worry about the open/closed distinction—namely, that if we cut an object in half, one piece will be closed and the other will not (and there is no principled way of saying which

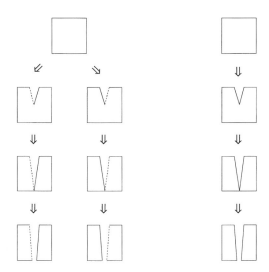

Figure 5.1
Wrong (left) and right (right) topological models of cutting.

will and which will not).[15] Suppose we dissect a solid sphere made of some perfectly homogeneous prime matter. Which of the two resulting half-spheres will be closed? This is an embarrassing question. But it arises, we submit, only on the basis of an incorrect model of what happens topologically when a process of cutting takes place. Topologically, the cutting of an object is no bloodstained process—there is no question of which of the two severed halves keeps the boundary, leaving the other open and bleeding (as it were). Rather, topologically the explanation is simply that the outer surface of the sphere is progressively deformed until the sphere separates into two halves. To put it differently, the cutting does not "bring to light" new surfaces that were trapped inside the sphere.[16] Rather, the model is that of a splitting oil drop. The drop grows longer and, as it grows, the middle part shrinks and gets thinner and thinner. Eventually the right and left portions split, and we have two drops, *each with its own complete boundary*. A long, continuous process suddenly results in an abrupt topological change. There was one drop; now there are two. And so in the case of the dissected sphere. There was one surface, and now there are two.

Of course, there is something deeply problematic about the magic moment of separation. However, this is true of every topological change. Consider:

— Two drops of oil move toward each other until they come into contact (the mirror image of the phenomenon of separation). A topological catastrophe takes place, and the topology of the overall configuration is suddenly altered. Two surfaces merge. Two drops become one.

— You press a stick into a ball of plasticine until the stick—*mirabile dictu*—breaks through to the other side. Once again we have a topological catastrophe taking place at the termination of a gradual, continuous process: a ball becomes a doughnut; the topology of the object undergoes an abrupt, qualitative change.

— A mushy blob starts growing a "finger" somewhere. The finger continues its growth until it eventually comes round to meet the main body again, forming a sort of handle. At the instant that it does so, the topology of the object changes: we had a sphere; now we have a torus.

Admittedly, stories like these all involve some topological magic.[17] *But nothing depends specifically on the boundary issue.* Every case of topological change marks a point where common sense reaches the limits of its theoretical competence, and it reaches that limit insofar as we are dealing with the ideal domain of dense, homogeneous bodies. The puzzle of the splitting object—we submit—is just a sign of this general fact. It can only disappear on a more complete assessment, which mereotopology simply cannot deliver. (This requires, we suppose, a step into the territory of qualitative kinematics.[18])

This line of reasoning applies also in relation to other puzzles. For example, Dean Zimmerman has a nice argument to the effect that the open/closed distinction seems to yield an implausible world in which

a certain class of objects [the closed ones] are unaccountably deferential to one another—always just managing to step out of one another's way—while they bang heedlessly into the members of another class of objects [the open ones]. Surely repulsive forces would *have* to be posited to explain such behavior. . . . But why should a certain shape of extended object be necessarily such that objects of that shape possess special repulsive powers? (1996b: 12)

The idea here is that we may imagine the same experiment performed twice. First we take an open cube and push it toward a closed cube with sufficient force so that they come into contact in 2 seconds. Then we do the same with two closed cubes. What reason can we offer to explain why in the latter case the two cubes will *not* come into contact in 2 seconds? Is their progression slower, or does it end sooner? We concede these are embarrassing questions, for "we seem forced to attribute repulsive powers of some kind to the cubes—an ability each cube has to "let the other know" that it has a skin of simples so that, if both the approaching surfaces are closed, the bodies can make sure to slow down or stop" (ibid.).[19] But there are other possibilities too. For instance, perhaps the two closed cubes *will* indeed come into contact. From the fact that two closed entities cannot *be* in contact it does not follow that they cannot *come into* contact, just as from the fact that two parts are connected it does not follow that they cannot be separated. But the coming into contact (just as the separation) determines a true topological catastrophe: there is a breaking through the relevant boundary parts and the two objects become one. (Think again of the two oil drops.) The two processes are dual: fusing is the reverse of splitting. And both involve a seemingly magic moment which runs afoul of the confines of pure mereotopological thinking.

5.7 Static vs. Dynamic Demarcations

As long as we confine ourselves to a topological model, then, cutting a solid object does not bring surfaces to light. Thus, the demarcation puzzle does not arise even in the presence of the open/closed distinction, and this blocks the argument against boundaries. We do not know exactly how this intuitive account can be extended to other forms of dissection—breakage, for instance. However, a complete picture is not necessary here. All we wish to emphasize is that the assumption that dissection will always leave two parts, one of which is closed while the other is open, may be reasonably challenged in cases where it would seem to yield unreasonable results. Other cases may be less clear, but so are intuitions. There is a complicated kinematic story to tell in each and every case.

We still have Peirce's puzzle, though. And here the puzzle seems truly problematic, for in this case the demarcation is perfectly static. Take any entity x. Does the boundary of x (or any piece thereof) belong to x or to its complement? Does the boundary inherit the properties—for example, color properties—of x or of its complement? There is no kinematic story to tell here. But how can we answer these questions without selecting one or other candidate at random and thus contravening the principle of sufficient reason?

Some cases are simpler than others. For instance, we have seen that material bodies such as tables or stones are naturally regarded as the owners of their boundaries (their surfaces). Thus, where the complement meets an object of this sort, *it* will be open. (The object, in turn, will be the closed complement of the complement.) We may also argue that immaterial bodies such as holes are not the owners of their boundaries, which belong to the material bodies that host the holes. Thus, where the two meet, the complement (host) is closed and the entity (hole) open. (The hole, in turn, is part of the host's open complement.) But even such simple cases immediately give rise to some dilemmas. For consider a typical hole that goes through a brick—a perforation. It is in contact with the brick; but there are also some regions of its boundary—corresponding to the entrance and the exit of the hole—that are not so in contact. (Hayes (1985b: 79) calls them 'portals'.) They are free. Or, if you prefer, they are in contact with the immaterial, airy body that constitutes the complement of the sum of the hole plus the sphere. The question is: Where do we place the boundary corresponding to those regions? With the hole? With the complement? Both answers seem bizarre. And either choice would seem to be thoroughly arbitrary.

We think we may acknowledge that such dilemmas are real, and yet insist on our friendly attitude towards boundaries. The actual ownership of a boundary is not an issue that a mereotopological theory must be able to settle. The theory only needs to explain what it means for two things to be connected. It does not need to give a full explanation of the underlying metaphysical grounds. After all, whether the boundary between hole and complement belongs to the hole or to the complement is a question to be answered by a theory of holes, not by a general theory of boundaries. (More generally, the theory of holes must answer the preliminary question of whether there is any boundary at all between hole and complement. Our own view is that the hole is a proper, undetached part of the object's complement, separated from the rest by a mere conceptual boundary—a notion to be clarified below.) By the same token, we can say that every instance of Peirce's puzzle—and of its temporal analogues, such as Aristotle's puzzle about rest and motion—is truly problematic and yet extrinsic to our present concerns. Give us a theory of black spots

and of white surroundings, and make sure to tell us who gets the boundary—the spot or the background. Give us a theory of events, and make
sure to tell us which gets the boundary—the state of movement or the
state of rest. If we accept this general line of response, we have a way of
disposing of the puzzle in its general form, or at least to dispel the clouds
of suspicion that surround it.

There is yet one last problem. Consider again the cutting of a solid
sphere in half. We argued that this process does not bring to light a new
surface. But, of course, we can *conceptualize* a new, potential surface right
there where the cut *would* be. In fact, we can conceptualize as many
boundaries as we like. As Smith (1994, 1995b) has pointed out, we often
make reference to purely imaginary, "fiat" boundaries of this sort, even
in the absence of any corresponding spatial discontinuity or qualitative
heterogeneity among the parts.[20] Bill's waist, the equator, the border of
your postal district, the Mason-Dixon line separating Maryland and
Pennsylvania —people draw imaginary boundaries where and when they
like. Is this not enough to give rise to the demarcation puzzle? How are
these boundaries to be allocated? There is *no fact of the matter* that can
support the ownership of a boundary such as the equator by one hemisphere rather than the other. Hence we cannot defer the solution to a
theory of the extended entities at issue, as we did above. The boundary
demarcating the Northern and Southern hemispheres is not only hard to
assign to either hemisphere. It *cannot* be assigned, no matter what our
theory of the globe might look like. And we cannot simply say that it
belongs to neither half, treating both hemispheres as semi-open entities.
The two hemispheres use up the whole globe by definition—no boundary
can be left as a thin, unowned slice *between* them. Finally, we cannot avoid
the problem simply by relying on the "imaginary" character of the equator. Fiat boundaries are imaginary in that they do not demarcate what
they bound through any intrinsically privileged feature of the underlying
physical world. But one need not be a Platonist to agree with Frege on
their reality:

One calls the equator an imaginary line, but it would be wrong to call it a line that
has merely been thought up. It was not created by thought as the result of a
psychological process, but is only apprehended or grasped by thought. If its being
apprehended were a matter of its coming into being, then we could not say
anything positive about the equator for any time prior to this supposed coming
into being. (1884: 35)

We react to this problem as follows: First of all, it is true that fiat boundaries are in a sense potential in that they do not actually *separate* anything from anything—they do not mark any actual discontinuity. However, this is not to say that these potential boundaries can be *actualized*. To say so would once again amount to the same mistake that we saw in connection with the modelling of the cutting process. There is no way you can bring the equator to light by actually cutting the Earth in half: dissecting the Earth would give you two Earth-halves, *each enveloped by a closed connected surface*, in such a way that the equator itself is gone forever. To put it differently, fiat boundaries are not the boundaries that *would* envelop the interior parts to which they are associated in case those parts were brought to light by separating the remainder. Wherever you have fiat boundaries in a physical object, you might generate real, bona fide boundaries in the corresponding places. But the former never *turn* into the latter—at best they *leave room* for them.[21] When we conceptualize a fiat boundary x we are not moving to a possible world where x involves a genuine discontinuity, and where the trouble of its belongingness arises. A more adequate picture is that when we conceptualize x we move to a possible world where x is replaced by two counterparts, x_1 and x_2, each bounding one side of the object. The actual transition from one world to the other is a complicated kinematic story. But the topology is clear. Fiat boundaries are *place-holders* for physically salient, genuine boundaries, but they are not themselves boundaries of this sort, not even potentially.

Accordingly, we have at least two ways of dealing with the demarcation puzzle in the case of fiat boundaries. On the one hand, we may say that the process whereby the fiat boundary is determined (i.e., "apprehended" in Frege's terminology) does indeed involve a form of indeterminacy: when we draw the equator to demarcate the two hemispheres, we simply leave the question of the equator's belongingness (hence the open/ closed distinction) unsettled. We naturally do so because that question has no practical relevance. But precisely for this reason the indeterminacy is innocuous: it is pragmatic, perhaps semantic—not ontological. On the other hand, we may also say that drawing a fiat boundary involves drawing *two* lines, one for each side. On this account, each hemisphere has its own equator and thus qualifies as topologically closed; but since the locations of the two equators are perfectly identical, the two hemispheres still qualify as externally connected and we can still speak of the equator as of a single thing.[22]

These two accounts are mutually exclusive, but neither need be exhaustive. In some cases the indeterminacy explanation applies most naturally, especially if the fiat boundary is the result of a conceptual operation (as when an one conceptualizes a sphere as being made of two hemispheres). In other cases it is the coincidence explanation that is more natural, especially when it comes to the fiat borders of the social world: Think of two adjacent peoples drawing the boundary between their territories. (One derivative advantage of this ontological doubling up is that it leaves room for the possibility of asymmetric boundaries, boundaries that bound their objects in certain directions only and not in others. Think, for example, of the old boundary between the former East and West Germany, which was recognized as a boundary from one side only.) It is not the task of mereotopology to decide which of these two accounts applies in which cases. For the purpose of mereotopology, what matters is that in both cases the demarcation problem disappears.

5.8 Further Boundary Issues

To sum up, we have reckoned that accepting boundaries calls for an account of the apparently troublesome distinction between open and closed entities, but we have also found the distinction to be ultimately unproblematic from a mereotopological perspective. In particular, a full-blown boundary-based theory such as **GEMTC** involves no incongruity related to the boundary concept. We now conclude this chapter by briefly mentioning some further issues that arise in relation to this important concept.

(a) Touching. The first issue concerns the relation of touching as this may occur between adjacent closed entities. We have seen that in **GEMTC** there is no possibility of genuine external connection between such entities. This is reflected by such theorems as (5.22) and (5.23), which immediately imply

(5.24) $ECxy \rightarrow (Clx \rightarrow \neg Cly)$.

Thus, the book is on the table, but these two objects are simply not in touch if this is understood in terms of external connection. This is in agreement with physics and with ordinary topology. Can we also do justice to the common-sense intuition that somehow the book and the table *are* in touch—that nothing can be squeezed in between them?

The proper answer is that this is not a mereotopological problem, but a metric one. Generally speaking, two closed entities are in touch (in this sense) if the distance between them is sufficiently small, or perhaps arbitrarily small. There are, to be sure, some interesting mereotopological relations that may be considered in this respect. For instance, Vieu (1995) has a relational predicate 'weak contact' that holds when two objects, albeit disconnected, are "vanishingly close" to each other. It can be defined as

(5.25) $WCxy =_{df} \neg Cxy \land Cx(c(ny))$,

where 'n' is a "neighborhood" operator defined by

(5.26) $ny =_{df} \iota w(Pyw \land Opw \land \forall z(Pyz \land Opz \to Pwz))$.

Informally, x is in weak contact with y if the closure of every open z including y is connected to x. This relation does indeed capture the intuition that nothing can lie between two entities that are in touch, even when these are closed. However, for our general purposes (5.25) is obviously inadequate: if space is dense, 'WC' is bound to be empty. Metric relations cannot be squeezed into the realm of mereotopology without the help of strongly simplifying assumptions.

(b) Vagueness.   We have been leaving out of account boundaries commonly conceptualized in terms of fuzzy zones rather than of sharp lines. However, it might be argued that ordinary objects and events may indeed have fuzzy boundaries of some sort. Clouds, dunes, hinterlands, let alone the figures of an impressionist painting, all seem to elude the idealized notion of a bounded object presupposed in the foregoing. Likewise, we may well agree that the rotation of a sphere takes place exactly where the sphere is located. But where exactly is Mary's lecture taking place? What is the exact spatial location—at this point in time—of John's phone call to his daughter? Can such questions be accommodated within the framework of a boundary-based mereotopological theory such as **GEMTC**?[23]

Some take the notion of a fuzzy boundary in a strict, literal sense, as for instance Michael Tye has urged:

There is no line which sharply divides the matter composing [Mount] Everest from the matter outside it. Everest's boundaries are fuzzy. Some molecules are inside Everest and some molecules outside. But some have an indefinite status:

there is no objective, determinate fact of the matter about whether they are inside or outside. (1990: 535)

If this is correct, then one should presumably think of the basic mereo-topological relations ('P', 'C', or both) as vague relations, and the account given above would call for some radical rethinking.[24] On the other hand, one may also understand the fuzziness of ordinary boundaries to be merely a semantic fact. In David Lewis's words:

The reason why it's vague where the outback begins is not that there's this thing, the outback, with imprecise borders; rather there are many things, with different borders, and nobody has been fool enough to try to enforce a choice of one of them as the official referent of the word 'outback'. (1986: 212)

This is the account we favor.[25] There are plenty of objects out there—plenty of slightly distinct and yet precisely determinate aggre-gates of land molecules. And when we say 'Mount Everest' or 'the out-back', each one of a large variety of such aggregates—each with its own perfectly crisp mereotopological structure—has an equal claim to being the referent of that term. If we wish, we can add that it is ultimately the vagueness of the relevant sortal concept (the concept *mountain*, or *out-back*) that is responsible for the way in which the referent of our expres-sion is vaguely picked out. It certainly is not the stuff out there that is vague.

On this account, then, vagueness is no issue for mereotopology. It is at most an issue that arises in connection with the drawing of fiat boundaries. As with the case of the equator, there may be a some degree of indetermi-nacy when we speak of such boundaries. But the boundaries are not in and of themselves vague, so we need not fuzzify our mereotopology.

(c) Dependence.   Our last remarks concern what is arguably one of the most important features of boundaries, namely, their status as ontologi-cally dependent entities. We have been talking of boundaries as of entities to be included in the domain of reference of a formal theory, next to the extended entities that they bound. But we also want to go out of our way to make it clear that the former are somehow dependent upon the latter. The surface of the table cannot exist in isolation from the table. It cannot be removed from the table, just as you cannot remove a hole from its doughnut. Can this fundamental feature be accounted for in a theory such as **GEMTC**?

The dependence of a boundary on its host is a case of genuine ontological dependence, as especially Brentano has emphasized.[26] It is not
merely a case of conceptual or *de dicto* dependence, as when we say that
there cannot exist a husband without a wife. Every husband, i.e., every
man who is in fact married, could have been a bachelor (or so we may
suppose.) But the surface of a table can only exist as a *surface of* a
table—perhaps only as a surface of *that* table. On the Whitehead-Clarke
approach, this relation of dependence is captured by the very fact that
boundaries are *secondary* entities: insofar as they are logically construed
out of more fundamental entities, they depend (i.e., supervene) on those
entities. In a boundary-based mereotopology, however, we need to give a
more direct account.

One general way of doing so is to assume suitable axioms that further
constrain the logic of the boundary operator 'b' and, more generally, of
the boundary relation 'B'. For instance, Smith (1993) has suggested a
postulate to the effect that every self-connected boundary is such that we
can find a self-connected entity which it bounds and which is not itself a
boundary, i.e., has an interior. In **GEMTC** this can be expressed as follows:

(D.1) $(SCx \land \exists yBxy) \to \exists y(SCy \land BPxy \land \exists zIPPzy)$.

(One needs the requisite of self-connectedness, for otherwise (D.1) would
be trivially satisfied by taking z open and setting y equal to $x+z$.) A
related postulate could be suggested to rule out outgrowing boundary
"hairs" or filamentous boundaries connecting separate voluminous parts:

(D.2) $Px(by) \to Px(c(iy))$.

In ordinary topological terminology, this amounts to requiring that a
boundary be always associated with a "regular" object. Other postulates
could be added along these lines, though of course one would eventually
have to resort to modal notions to fully do justice to the sort of *de re*
necessity involved in the relation of ontological dependence.[27]

We may also consider a different way of dealing with this relation. We
could think of a theory of dependence as a chapter of its own in the
overall theory of parts and wholes (as suggested by Husserl in the *Logical
Investigations*), and we could then add suitable postulates to capture the
dependent status of boundaries (along, perhaps, with the dependent
status of other topologically salient entities, such as holes). Formally this

could be done by supplementing 'P' and 'C' with a new primitive predicate, say 'D', reading 'Dxy' as 'x is dependent on y'. For instance, Fine (1995a) suggests the following set of axioms:

(D.3) $Dxy \wedge Dyz \rightarrow Dxz$

(D.4) $Pxy \rightarrow Dyx$

(D.5) $\exists y(Dxy \wedge \forall z(Dxz \rightarrow Pzy))$.

(D.3) would be part of the ground theory of dependence, whereas (D.4) and (D.5) would serve the purpose of linking mereology and dependence. One could then address the dependent nature of boundaries by adding a specific principle linking topology and dependence:

(D.6) $Bxy \rightarrow Dxy$.

Although more involved, such an account would be very much in the spirit of the general strategy we defended in chapter 4—start with mereology, and then add topology and other theory fragments.

6 Parts and Counterparts

There is another important issue left open from our discussion of part-whole structures in chapters 3 and 4, and it concerns the nature of proper parts. Mereotopology treats all parts as objects of quantification—full-fledged entities endowed with the same right to existence as the wholes to which they belong. But such a position needs scrutiny. It goes without saying that the mereological sum of this board and that wine glass has two objects as parts: the board and the glass. Much less clear is the status of, say, the left and right halves of the board, or of the stem and the rest of the glass. Such parts seem much less conspicuous and are of lesser cognitive salience. (Cognition does not seem to find them, but to create them.[1]) This is because portion of their boundary is of the *fiat* sort: the result of an imaginary demarcation.

So uncospicuous these demarcations can be, undetached parts might as well be treated as potential objects only.[2] How exactly can we account for this intuition?

6.1 Two Aspects of Potentiality

There are two distinct ingredients in the idea that certain proper parts are merely potential objects. The first is *detachability*. When we think of an object that is topologically all of a piece, we neglect its parts. It is not that we positively think that such an object has no parts. But we do not take parts into account until there is some need to detach—physically, or in the imagination—one or more of them.

Consider a square board. We may cut it in half by sawing along a north-to-south line, or we may cut along an east-to-west line. In the first case, the east and west halves of the board would be separated. In the other case, the north and south halves would be separated. Suppose we in fact cut in the second way, so that we now have the north and south halves of the board in front of us, well apart from each other (figure 6.1). Clearly, the east and west halves are not two particularly good candidates

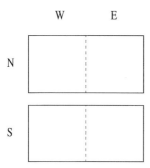

Figure 6.1
A board, its north and south halves, and its east and west halves.

for objecthood—they are not unitary, self-connected wholes, and neither are they completely demarcated from their environment. This at least is cognitively plausible: the unity of, say, the north half cognitively dominates over the double handicap of, say, the east half, which is spatially scattered and whose scattered parts are undetached from the corresponding scattered parts of the west half. This is not to say that the east half is not available for reference and quantification: we *are* talking about it, and our mereological theory has a place for it in its domain of discourse. (This was the point of the general Fusion Axiom (P.8).) Rather, the east half does not seem to pass the test for genuine objecthood. It is overwhelmed by other objects, such as the north half. And if the east half does not pass the test after the cut—one could argue—why should we be tempted to say that it did before? If it was an object, it was only a potential object.

Detachability, then, is one crucial ingredient of the notion of potentiality that arises in relation to proper undetached parts. The other fundamental ingredient has to do with *counting*. When we think of the board, it is as if we assigned it a number. There is *one* thing there, one maximally self-connected entity: not two, or three, or some other number corresponding to some other way of partitioning the board. In order to assign the board the number two, say, we should imagine that it was cut in half—or we should cut it in half, thereby losing the unity of the board. Every partition of an object determines a count, but as every partition excludes any other, so every count excludes any other. Thus, a way to reconcile the rigidity of the actual counting with the possibility of multiple partitions is to say that any counting that assigns a value greater than one to a maximally self-connected object must be a counting of potential parts.

Always in the background of our part-whole thinking is thus a deep tension. We want to talk about parts in the same way in which we talk about whole objects. We want to quantify over parts, compare them, describe them explicitly or implicitly. (Well before the cut we can say that the north half of the board is as big as the south half; that it lies above it; that it is rectangular.) However, in many cases we also want to make sure we are talking about things that are purely potential—parts that are not individuated except by reference to the wholes to which they are attached. In this sense, the undetached north half of the board is not an object to be included in an inventory of the world *over and above* the board itself, even though both of them are included in the domain of our part-whole theory. How is this tension to be explained? Can we reconcile these two seemingly opposite attitudes towards the status of parts?

As we have seen, we go here beyond a purely mereological concern, and topological notions are involved in a crucial way. No quarrel arises about the nature of those parts that already qualify as ordinary objects. More generally, detached parts—i.e., parts that are maximally self-connected—do not pose any problem. Take the scattered whole consisting of the board and the wine glass. The board and the glass are parts of the whole—they are proper parts. Yet surely they pass any test for objecthood (and do so even if the whole does not). Rather, the quarrel concerns the status of such things as the north half of the board, or the top half of the glass. The problem arises in relation to a whole's *undetached* parts—parts that do not have that full-fledged thingy character that distinguishes maximally self-connected entities. Does their potential status call for a reinterpretation of mereotopology?

6.2 The Idea of a Countermereology

The notion of a potential object is a modal notion. To say that the north half of the board is potential is to say that it counts as an object only in some possible world: it does not in fact pass the test for objecthood, but it could.

Now, it is difficult to express this suggestion in terms of standard modalities, i.e., in ordinary modal-logic terms. This is because we want to deal simultaneously with entities from different possible worlds—we want to quantify simultaneously over whole objects as well as over their parts—and this can hardly be done within an ordinary modal-logic setting.[3] However, we can also express modal intuitions without departing from the standard apparatus of first-order logic: if we include possible worlds in the domain of quantification (perhaps along with a suitable accessibility relation), we can express modal facts as quantified statements about possible worlds (necessity amounting to universal quantification and possibility to existential quantification). Nothing then prevents us from speaking simultaneously about parts and wholes even when these belong to different worlds.

One way of pursuing this line would be to treat the mereotopological primitives as involving an additional argument place.[4] We would have a three-place predicate for parthood, reading '$Pxyw$' as 'x is part of y in world w', and a three-place predicate for connection, reading '$Cxyw$' as 'x is connected to y in world w'. Defined predicates could be handled accordingly—for instance:

(6.1) $Oxyw =_{df} \neg \exists z(Pzxw \wedge Pzyw)$

(6.2) $ECxyw =_{df} Cxyw \wedge \neg Oxyw.$

The idea that a proper undetached part x is a potential object could then be expressed by means of a mereotopological statement to the effect that x is separable from any given adjacent part z, i.e., that there is at least one possible world in which x is disconnected from z:

(6.3) $Pxyw \wedge Pzyw \wedge ECxzw \rightarrow \exists w' \neg Cxzw'.$

More emphatically, we could assert that there is a possible world w' in which x is maximally self-connected, this notion being defined along the lines of the predicate 'MSSC' (see (4.24)), but in terms of the new ternary primitives.

This is a viable account, but it still has one major problem: it calls for a radical rethinking of the background mereotopological machinery. And this is not just heavy but straightforward labor: treating parthood and connection as world-sensitive relations requires a thorough examination of many deep questions. Should identity too be relativised to possible worlds? Can one and the same object have different parts in different possible worlds? Can it be connected to different things? And what mereotopological story can we tell about cross-wordly entities such as the

sum of your hands and Pegasus's wings, or the sum of that thing which is
your right hand in this world w and that thing which is your right hand in
a different world w'? Some of these questions run very deep indeed.

But here is a different account. Let us keep 'P' and 'C' as they are, but
let us express modal intuitions using the counterpart idiom of Lewis 1968.
According to counterpart theory, modal facts concerning an entity existing in this world are non-modal facts concerning entities existing in some
(accessible) possible worlds. For example, the fact that John could be
shorter than he actually is is the fact that there is a possible world in which
there can be found a certain individual corresponding to John in some
relevant sense—John's *counterpart*—who *is* shorter than John. John's
counterparts are men John would have been had the world been otherwise, and to say that John could be shorter than he actually is is to say that
some such counterpart is shorter than John. Now, we can deal with the
idea of potential parts in a similar fashion. Potential parts, we may say, are
partial counterparts of the individual wholes to which they belong. They
inhabit different worlds. And the conditions for the truth in a world w of
a mereotopological statement

x is a proper undetached part of y

can be given by

in some world w' (accessible from w) there exists an object, x, that is a
partial counterpart of y.

This means: there is an object, x, that is a counterpart of y, but not a
complete counterpart; for w' is chosen among those worlds in which there
simply is no such thing as a complete counterpart of x. For instance, the
conditions for the truth in a world w of the statement that North (the
north half of the board) is a proper part of the board are given by the fact
that there is some accessible world w' in which North is a partial counterpart of the board. In this way not only can we deal with modal facts about
parts without departing from the standard apparatus of first-order logic
(by including worlds in the domain of quantification). We can do so
without departing from our basic mereotopological apparatus either. The
mereotopology stays, and to treat parts as potential is simply to treat
them as belonging to different possible worlds than the wholes to which
they are mereologically related.

The problem of *which* thing or things, in an accessible world w', are partial counterparts of a given actual object is not a major problem. One has a similar difficulty with counterparts in general. Which one is John's counterpart in a world where someone looks exactly like John except for the fact that his right eye is red, and somebody else looks exactly like John except for the fact that his left eye is red? Let us just say that it is a crude metaphysical fact that some objects, and not others, are John's counterparts. Likewise it is a crude metaphysical fact that some objects, and not others, are John's partial counterparts. The important fact is that only those things may qualify as partial counterparts of John that are parts of John. (Note: the partial counterparts of John in a world w' are not those objects of w' that are the counterparts of John's parts. They are exactly John's parts.)

Here one might object that the notion of partial counterpart does not exempt one from recognizing the primacy of the notion of part. But this problem too is immaterial from the point of view under consideration. Grant the primacy of parthood: we can make out the partial counterparts of an object x in a neighboring world w' only to the extent that those objects are parts of x in the actual world. But from this one cannot infer much about the nature of the parthood relation. The metaphysical priority of the partial counterpart relation over the parthood relation is not challenged: the burden of the above-mentioned truth conditions would still be entirely on partial counterparts.

Prima facie, then, the idea of a framework for revisiting parthood in terms of partial counterparthood—a *countermereology*, to give it a name—is not incoherent. Let us look deeper into it. In a sufficiently articulated mereotopology such as **GEMTC**, the crucial notion of an undetached part (UP) can easily be characterized: an undetached part of a given whole y is any proper part of y that is not maximally strongly self-connected:

(6.4) $\text{UP}xy =_{df} \text{PP}xy \wedge \neg\text{MSSC}x$.

This characterization is actually somewhat oversimplified, for we should rely on the parametrized predicate 'ϕ-MSSC' to avoid undesired models (see (4.24)). However, (6.4) will serve well our purposes for the moment.

Let us now focus on those parts that are undetached parts of self-connected wholes—call them ordinary parts (OP). This is necessary if we want to make sure that we are talking about the right sort of things—e.g., North as part of the board rather than part of the sum board+glass:

(6.5) $\text{OP}xy =_{df} \text{UP}xy \wedge \text{MSSC}y$.

Clearly, this predicate is irreflexive and asymmetric, or, more generally, intransitive:

(6.6) $\neg \text{OP}xx$

(6.7) $\text{OP}xy \rightarrow \neg \text{OP}yz$.

Moreover, it follows from (6.5) that nothing can be an ordinary part of more than one object:

(6.8) $\text{OP}xy \wedge \text{OP}xz \rightarrow y=z$.

A countermereology can then be characterized simply as the result of treating this predicate as expressing partial counterparthood. And this can be done by supplying suitable axioms for 'OP' in the general spirit of counterpart theory. Where 'I' is the relation of 'being in' holding between objects and worlds, the following set of axioms appears to be a good first step in this direction:

(CP.1) $\text{OP}xy \rightarrow \exists w \text{I}yw$

(CP.2) $\text{OP}xy \rightarrow \exists w \text{I}xw$

(CP.3) $\text{OP}xy \wedge \text{I}xw \rightarrow \neg \text{I}yw$

(CP.4) $\text{OP}xy \wedge \text{OP}zy \wedge \text{I}xw \wedge \text{I}zw \wedge \text{C}xz \rightarrow x=z$.

On the intended interpretation, an object x is in a world w if it passes the test for objecthood in w. Accordingly, (CP.1) and (CP.2) assert that only things that pass the test for objecthood in some world have or are ordinary parts: every ordinary part must be part of an object which is in some world, and the part itself must be in some world (these axioms make 'OP' analogous to Lewis's counterpart relation). (CP.3) ensures that no ordinary part of an object can be in the same world as that object. This could actually be strengthened to

(CP.3′) $\text{I}xw \wedge \text{I}yw \wedge \text{PP}xy \rightarrow \neg \text{SC}y$, which asserts that two distinct individuals which are one part of the other can be in the same world only if the latter is a scattered object. Finally (CP.4) asserts that no two connected ordinary parts can be in the same world: if both are ordinary parts, then they can only be connected if they properly overlap or if they are externally connected, and in both cases they fail to pass the test for objecthood.

We can be flexible here concerning the exact properties of 'I'. In Lewis's counterpart theory, 'I' can itself be interpreted as parthood: being *in* a world is being *part of* that world (see Lewis 1983, Postscript A). In the present context this can only be confusing. All we shall say is that wolds are not mereologically contained in their inhabitants, regardless of whether the containment holds in the opposite direction:

(CP.5) $Ixw \rightarrow \neg Pwx$.

6.3 Some Questions Answered

Is countermereology a viable account of the feeling that proper undetached parts are potential?

Consider first (CP.1) and (CP.2). These two axioms are meant to capture the intuition that we are taking ordinary parthood as a form of counterparthood, which is a relation holding between individuals. In counterpart theory, worlds are included in the range of individual variables; but this is not to say they are individuals on a par with the things that inhabit them. This is explicitly embedded in Lewis's theory via two axioms of which (CP.1) and (CP.2) are the analogues in terms of 'OP'. On the other hand, it may be objected that in mereology *everything* is an individual of the same sort as anything else. Mereologically speaking, the universe—the fusion of everything, possible and actual—is an individual on a par with the smallest cells that compose our bodies. At least, this is true if we accept the unrestricted Fusion Axiom (P.8). If so, then not every individual can be *in* some world, and not every individual woud have all of its parts actualized in some world or other. (There are also some bizarre individuals that span across different worlds, such as the cross-world mereological fusion of an object's counterparts.) However, this seeming dilemma arguably disappears as soon as we take seriously the underlying topological requirements. The sort of whole that can have ordinary parts must be topologically connected, and it is plausible to suppose that such a requirement can only be satisfied by objects located in a single world. This is the point of (CP.1).

Consider now (CP.2) and (CP.3). Together, these axioms capture the basic idea that potential parts are *actualized* in some worlds, where they are isolated from the rest of their whole. Ordinary parts are potential in the precise sense of existing as maximal wholes in some possible world (distinct from the actual world). However, it may be objected that this overlooks the possibility of parts that are, not only undetached, but undetachable. The distinction goes back to Husserl's distinction between independent and non-independent parts,[5] even though the cases we have in mind are not, as in Husserl, cases of "abstract" parts (such as the visual quality or the extension of a given figure). Just suppose that certain parts of certain objects are simply not such as to ever exist alone, in isolation from those objects. For instance, we saw in section 5.8 that boundary parts (such as the surface of a material object) are ontologically dependent on the things that they bound. On some understanding, this means that they are unseparable from such things. There is no possible world in which you have the surface of the board separate from the interior.[6] Now, for such parts one cannot say that their non-actuality is to be explained in terms of partial counterpathood. So, for such parts, the account does not apply.

We think this need not be a problem for countermereology. Dependent parts are not potential parts. But neither are they actual parts, if actuality is taken to imply autonomous individual existence. Dependent parts are parasitic on the wholes to which they belong. And the parasitic nature of their existence is enough to do justice to the intuition that they are not on a par with ordinary wholes. Countermereology, one could say, explains one half of the story; the other half is explained by the theory of dependence.

(CP.3) also reflects the crucial intuition that ordinary parts are not actual, if the whole is actual. The weakening obtained by dropping this axiom would be germane to the view that at least some ordinary parts can be actual. For instance, van Inwagen (1981) argues against the acceptance of *arbitrary* undetached parts but concedes that some undetached parts (e.g., the cells of a living organism) be counted as actual material beings.[7] Although this would be in the spirit of greater generality, we see (CP.3) as a necessary ingredient of countermereology. After all we are working with a purely mereotopological characterization of the notion of an ordinary part: from this perspective any putative counterexample to (CP.3) seems to involve a departure from that characterization and to incorporate into mereotopology principles that go well beyond it (e.g., in van Inwagen's case, biological principles). If a proper part qualifies as actual, it is because it is fully individuated by a boundary of some sort—and van Inwagen's cells are no exception.

Finally, consider (CP.4). This axiom prevents any two overlapping ordinary parts from being in the same world. But it also prevents two *adjacent* ordinary parts from being in the same world. This might sound contrary to the intuition that an ordinary part is a potential entity insofar as there exists a possible world in which it is detached from the other parts. The north half of the board is potential insofar as it can be separated from the south part; but why should we force these two parts to be in *different* worlds? The answer is simply that countermereology is not a mereology for counterfactual reasoning. Of course there are worlds where the mereology is preserved and only the topology is altered—worlds in which the two parts coexist, but are separate. This is the sort of world that can enter the three-place parthood and connection relations discussed at the beginning of section 6.2. But this is not the sort of world that we are interested in at this point. In countermereology, worlds are selective devices. For an ordinary part to be in a world is for it to count as an object, i.e., to be singled out and isolated from the rest. But since we are leaving the mereotopology fixed, this isolation cannot be the result of a true separation. It is purely cognitive—the result of a purely imaginary demarcation (a fiat boundary).

It may also be worth emphasizing that (CP.4) prevents any two distinct individuals from the same world to properly overlap unless at least one of them is topologically scattered. This is because if x and y properly overlap, they are connected, and if they are self-connected they also qualify as ordinary parts of their sum $x+y$, or of some maximally connected extension z of $x+y$. Hence by (CP.4) they must be in different worlds:

(6.9) $POyz \land Iyw \rightarrow \neg Izw$.

This might sound exotic and hard to picture. However, there is a simple explanation for what is going on here. In countermereology, to say that x and y properly overlap is to say that they are in different worlds and that there is a further world (distinct from the previous two by CP.3) contain-

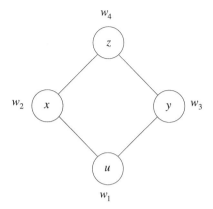

Figure 6.2
The countermereology of overlap and underlap.

ing an entity z (the maximally connected extension of the sum $y+z$) of which they are partial counterparts. In addition, it also happens that the common part of these two objects, u, is itself a partial counterpart of z. And by (CP.4) u must be in a further world. The network of relations is depicted in figure 6.2. There is some formal complexity, but this complexity is just what we expect from the part-whole analysis of the situation. And this point seems generalizable: in spite of the modal machinery, countermereology is not incompatible with the customary way of treating mereological relations and operations.

6.4 Mereological Relativity

One more general remark is in order. It concerns the notion of maximal connectedness involved in the notion of an undetached part introduced in (6.4), on which the entire formulation of countermereology hinges. As noted, we have relied on the predicate 'MSSC' but a more general account would require the parametrized predicate 'ϕ-MSSC' (where ϕ is any property or condition; see (4.25) for the definition). The reason for this generalization is that the notion of an undetached part is really a relative notion: the board is a maximally connected *material object*, but is an undetached proper part of any whole consisting of the board itself together with an adjacent piece of the complement.

It is not difficult to take this need into account. First, we can replace (6.4) by the parametrized definiton schema:

(6.4′) ϕ-UP$xy =_{df}$ PP$xy \land \phi x \land \neg\phi$-MSSC$x$.

This makes it possible to say, for instance, that if ϕ expresses the property of being a material object, then the north half of the board is a ϕ-UP of the board, but the board is not a ϕ-UP of the hybrid sum consisting of the board plus the piece of complement. Next, definition (6.5) can be emended in a similar fashion:

(6.5′) ϕ-OP$xy =_{df} \phi$-UP$xy \land \phi$-MSSCy.

And, finally, the countermereological axioms can be rewritten accordingly, as a set of axiom schemata involving the relation ϕ-OP for arbitrary predicates or conditions ϕ. Admittedly, the upshot will be a much more cumbersome theory. But, once again, complexity is not the issue here.

Here is an alternative way of looking at the relativity of the notion of ordinary part. We are trying to explain a natural tension that arises from the fact that, on the one hand, mereology quantifies over all parts, and on the other hand, some parts do not have the thingy character that we expect from ordinary objects. In relying on the topological notion of maximal connectedness (MSSC), we are taking topology to supply the relevant conceptual distinctions—in particular, the distinction between an ordinary undetached part and a fully demarcated, detached part. However, we have seen in chapter 2 that topological unity is only one of various kinds of unity that can be appealed to in singling out what Husserl called the "pregnant concept of a whole." Homogeneity, causal unity, functional unity, teleological unity are all quite relevant. This is nicely exemplified by a passage by Fritz Heider:

A chair consists of many parts. How is it possible that it is nevertheless a unit? . . . It would hardly be feasible to apprehend one portion of the chair and a part of the adjacent air as a single object. . . . There is a dependence between the parts of the chair which does not exist beween the parts of the atmosphere. (1959: 8)

This variety of pregnant factors is lost in the basic countermereological framework based on (6.5). And since everything is connected to its complement, such a loss may have disastrous consequences for the whole account. (Nothing would qualify as a maximally self-connected whole except for the entire universe, so 'OPxy' would only be true when y is the universe.) A most general way out would be to rely on a variety of predicates besides 'MSSC' to define a corresponding variety of notions of undetached part. The relativization of 'OP' afforded by (6.5') is one way to make the account more flexible at least with regard to the topological dimension. A object may qualify as a complete whole with respect to a certain condition φ but not relative to others. Depending on which conditions we consider, it may or may not be included in a *bona fide* inventory of the world.

From this perspective, countermereology is but one instance of a more general view concerning the nature of parts and wholes. It is the view that *only some entities, among those countenanced by mereology, must be included in an inventory of the world*. Unlike set theory, mereology does not draw any distinction between the ontological category of the first and second arguments of its primitive relation: both arguments of 'part of' are individuals, entities of the lowest logical type. This is why we can talk about parts in the same way in which we talk about whole objects. However, this is not to say that such entities are all on a par. We may choose among various ways of drawing up an inventory of the world; we can decide to include the parts, or we can decide to include the whole. *But once we have chosen there is no room for double counting.* And this explains the tension mentioned above. If we include a part, we cannot include the whole; and if we include the whole, we cannot include the parts.

Call this the Minimalist View. The Minimalist View then says that once we have fixed a domain by means of our mereotopological theory, we can draw up an inventory of a world *w* by listing some elements in the domain of *w*, either individually or collectively, in such a way that in the end each and every element is either included in the list or overlaps something included in the list. More precisely, this can be expressed in terms of a twofold requirement. On the one hand, the Minimalist View says that every admissible way of drawing up an inventory must satisfy a non-re-dundancy condition: If *x* properly overlaps *y* and *y* is included in the inventory, then *x* is not itself to be included. This avoids double counting. On the other hand, the Minimalist View also says that one should not leave anything out of the inventory. So the converse constraint must also be met: everything in the mereological domain must overlap something that is included in the inventory. (We are relying here on the fact that the underlying mereology is extensional, so that if *x* properly overlaps *y*, then

x can be split into two parts, one included in y, and one disjoint from y.) Putting the two constraints together, we can therefore take the Minimalist View to be characterized by the following biconditional:

(M) x is included in an inventory of a world w if and only if x does not overlap (at the time when the inventory is drawn up) any distinct y that is itself included in the inventory.

This is of course not a definition of what it takes to be included in an inventory of the world (on pain of circularity). But we may take (M) as a postulate characterizing the view under consideration: any inventory of the world must satisfy (M).[8] Countermereology may then be seen as a way of implementing one particular instance of (M): the one that is obtained by including in an inventory only things completely demarcated by a genuine (non-fiat) boundary. We may call this the Normal View:

(N) x is included in an inventory of the world only if x is strongly maximally self-connected (or strongly maximally self-connected relative to some condition ϕ).

6.5 Counting Policies

One can at this point imagine stronger or weaker conditions corresponding to stricter or looser count policies. For instance, the strictest count policy is atomistic—never count an entity that has proper parts:

(S) x is included in an inventory of the world only if x is mereologically atomic, i.e., has no proper parts.

Of course, this presupposes that the domain be fixed by an atomistic mereology, so that everything is ultimately constituted by mereological atoms. (The converse of (S) is also true on the Minimalist View, since it follows from the right-to-left condition of (M) using (S). So this policy can actually be formulated by requiring an inventory of the world to consist of all and only the mereological atoms in the domain.)

By contrast, the loosest possible count is monistic: the only entity to be included in an inventory of the world (the only entity that exists in the ordinary sense) is the total sum of everything included in the domain. More generally this can be expressed by the following condition:

(L) x is included in an inventory of the world only if x is mereologically maximal, i.e., not a proper part of anything.

But note that if the domain is not closed under unrestricted mereological
fusions, (M) may be inconsistent with (L). (Think of a domain including
x_1, x_2, and x_3 along with the sums x_1+x_2 and x_1+x_3, but not $x_1+x_2+x_3$. If,
say, x_1+x_2 is included in the inventory, x_1+x_3 cannot be included (owing
to the non-redundancy condition of (M)), so x_3 must be included (owing
to the exhaustivity condition of (M)), contrary to (L).)

Are there any worthy policies between these two extreme cases, apart
from the natural count policy (N) of countermereology? There must be,
since (N) is in itself a rather extreme position, ruling out the possibility
that any proper undetached part be ever included in an inventory of the
world. Van Inwagen's position cited earlier is a potential candidate: only
some undetached parts exist:

I think that the cells living things are made of are . . . *unitary* things, things having
an entelechy; in this respect they are like the men, women, and dogs . . . of which
they are parts. (1981: 133)

This is an interesting view, but as it stands it is inconsistent with the
Minimalist View. Van Inwagen is willing to have men, women, and dogs
along with their cells, and that violates the non-redundancy constraint.
Indeed, Van Inwagen's view is best regarded, not as a moderate alternative to (N), but as an alternative to (M). According to van Inwagen (1987,
1990), the correct answer to what he labels 'The Composition Question'
is that certain xs compose something if and only if their activity constitutes a life. In our terminology, that means that a mereological aggregate
is to be included in an inventory of the world if and only if it is a living
organism. And this violates (M) to the extent that living organisms can be
mereologically included in other living organisms, as the cell example
suggests. We prefer to keep (M) and regard the cell case as indicating a
shift from one level of description to another—a shift from an inventory
of men, women, and dogs to a stricter inventory of cells and other micro-
organisms. Putting them all in the same inventory on account of their
different survival properties is disguised double counting, just like putting
the north half along with the whole board.

What intermediate options are left, then? One can go in both directions.
There are circumstances where a looser criterion than (N)—though not as

loose as (L)—seems appropriate; and there are circumstances where a stricter criterion—though not as strict as (S)—seems appropriate.

Looser counts seem appropriate when a scattered aggregate is counted as one on account of its being unitary in some relevant sense. You buy a six-pack of beer; the cashier reckons six items, but they count as one in your shopping list. The unifying factor may be purely cognitive, or it may be explained in terms of some sort of non-topological connectedness, as mentioned in section 2.2. A gravitational system or an electric circuit may be counted as one on account of their forming a closure system under certain functional relationships (exerting gravitational force, or transmitting energy). If the mereotopological structure of the domain of reference is as fine-grained as demanded by the physical sciences, physical objects of the garden variety can only be included in an inventory of the world as a result of such a loosening of (N).

Stricter counts also seem appropriate in various cases. The doughnut and the hole are sure to be connected—yet we may want to list them separately in an inventory of the world. Within the constraints set by (N), one could only do so by treating holes as categorially different from their physical hosts. That might be worth the price. But then the hole must also be listed separately from the rest of the doughnut's complement, and that can hardly be handled in terms of categorial distinctions. A hole is naturally regarded as a homogeneous and yet distinguished undetached proper part of the all-comprehensive complement of its material host, and we may have reasons to include such a part in the inventory of the world rather than the whole complement, contrary to (N). Or again, in the temporal realm, events may well be connected and yet such as to deserve separate consideration. Two things just happened; Mary kicked the ball and the ball started rolling. Nothing happened in between. And nothing separates one event from the other. Yet we may want to include the two of them separately in the inventory of the world (if this is to include events at all).[9] This is not double counting, so (M) is safe. But (N) is violated. More generally, (N) will be violated whenever we include things or events that are demarcated only by a fiat boundary, a boundary to which there corresponds no real physical discontinuity. (For another example, from the ontology of geography, two nations may be connected through a common border, as Canada and the United States are, but that doesn't deprive them from the unity needed to be counted separately. So if we are willing to put geographic entities in the inventory of the world, then again (N) is to be violated.)

All of these seem to us admissible exceptions to (N), pointing in the direction of much more sophisticated count policies. Counting as one, we may say, is a function of possessing a certain unity. But the relevant notion of unity may vary, and the general notion of φ-connectedness does not capture every interesting case. There are other senses of unity, and for these senses (N) will set the wrong standards.

7 Modes of Location

Now let us turn to the relation of spatial location. Besides mereology and topology, the relation between an entity and its place—the place that it occupies or where it is located—is a fundamental ingredient in a general theory of spatial representation. And it must be treated independently. We cannot treat it as mere identity, for we want to make sense of the possibility that the same place be visited by different things at different times. (We also want to allow for the possibility that it could host different visitors at the same time.) Nor can we treat location as a mereological relation: a moving object can hardly be described as undergoing mereological change. And even if the space occupied by a rolling stone (at some instant of time) shared parts with the stone, we would still have to single out those parts as special parts of some sort. Likewise for topology: we may want to say that a stone is always connected to the space that it occupies, but this would not account for the intimacy of this link. If the two are connected but do not overlap, they must be externally connected; yet this could hardly be understood as a case of connection between contiguous or continuous entities. Even if it turned out that location is dispensable due to some combination of mereology and topology, still we should not put the cart before the horses at this point. Methodological prudence suggests that we start by treating this relation as an independent primitive next to parthood and connection. A reduction might come as a theorem, but it cannot be an axiom.

7.1 Having an Address, Being in a Place

The intuitive ground for location is the notion of an address. Spatial entities are spatial insofar as they cannot but be somewhere. They must possess an address at which they can be found. Not only people, but by extension also material objects such as tables and chairs can be assigned an address in this sense. More elusive entities, too, such as the surface of the table and the hole in the doughnut, have an address. And so do events, elusive as their address may be. (Where exactly did Louise's logic test take place?)

Though elementary, the notion of an address involves intricacies that bear witness to the complexity of location. We can distinguish, to begin with, a *permanent* and a *temporary* (non-permanent) notion of address. John lives in Manhattan, and even if he drives somewhere outside Manhattan, he still retains an address there: that's his permanent address (the customary notion of an address). By contrast, when John flies to Paris or goes for a stroll in Central Park, his temporary address changes. He may retain his permanent address, but he is in fact in a different place than he was before the flight or before the stroll. This is not the ordinary notion of an address, but its use in certain contexts is quite appropriate. (Think of a taxi driver asking for John's address when John is calling from a restaurant.)

On a different line, we may further distinguish a spatially *minimal* and a *broad* (non-minimal) notion of an address. Every time John moves (including when he only moves, expands, or contracts parts of his body), his minimal address changes. This holds clearly in the temporary sense, and can be extended to the permanent sense. On the other hand, John may move around Manhattan and still be *in* Manhattan. His moving does not necessarily carry along a change in his address, if this is understood in a broad sense. (This may well be a matter of focus. Think of a Martian mentioning Earth as the position of John.)

A third classificatory criterion draws upon the semantic structure of the singular terms used for denoting addresses. *Spatially unstructured* terms include proper names of addresses or definite descriptions whose semantic complexity is not spatially relevant. Consider, respectively, 'John lives in Manhattan' and 'John lives in the Empire State Building', or 'John lives in the most expensive building ever built'. *Spatially structured* terms are either proper names or definite descriptions whose semantic components are overtly spatial. The interesting cases are provided here by coordinate systems or street numberings: 'John is at 40° North (of the Equator) and 74° West (of Greenwich)'; 'He is at 5th Avenue and 34th Street'; 'He is at 77 5th Avenue'. Address terms may also be used projectively. For instance, the temporary address of a flying plane at a certain point in time might be given in terms of the plane's projection on the surface of the Earth ("ground position"). These and similar cases, however, when the address name makes reference to a place that is not the place where the object is really located (either temporarily or permanently), need not be of our concern here. Also, for simplicity we ignore the case of indexical spatial expressions such as 'here' or 'three feet from there', or of any other expression that essentially involves a viewpoint, such as 'to your right' or 'behind the Empire State Building'.

Now, one understands addresses in the permanent sense in terms of the temporary notion of an address: to have a permanent address α implies having the possibility of temporarily being at α. Moreover, to systematize the temporary notion of an address—the address you have when you are where you are—it is convenient to rely on the notion of a minimal address. Your minimal address, in fact, is sure to be common to whatever temporary address you might have. Thus, *your present temporary minimal address gives your exact location at this moment of time*, the region of space presently taken up by your body. John's present temporary minimal address gives his exact location: not Manhattan, but the much smaller space carved out of the air or of whatever medium he might be in (water, if he is swimming; concrete, if he betrayed his godfather).

This notion of exact location is the notion we are interested in here. It is closely related to the idea of a boundary, for the exactness of an object's location is determined by the location of the object's boundaries.[1] And it is best appreciated when spatial regions are included in the *prima facie* ontology (although, as we mentioned, we wish to remain neutral with respect to their ultimate ontological status). Indeed, in a preliminary characterization the relation of exact location is a relation whose second argument can *only* be a region of space—a place. Even if we can talk of John's being located *in a column* in the basement of his godfather's house, this would not be the sense in which we are using the term here. Instead, John would be located in a space carved *inside* that column. On the other hand, the first term of the location relation can be whatever sort of entity we have in our spatial ontology—spatial regions included. We can speak of John's being located at a region r; but we can also speak of John's body, of the sum of his present intradermal events, or even of region r itself being located at r.

7.2 Principles of Location

In terms of exact location, assumed as a primitive relation, we can now unfold a more detailed picture of locative relations by exploiting the part-whole structure of spatial entities.[2] Let us use 'L' for this new primitive, reading 'Lxy' as 'x is exactly located at y'. With the help of mereology we can then expand the set of available locative relations to include cases of inexact location:

(7.1) $PLxy =_{df} \exists z(Pzx \wedge Lzy)$ (Partial Location)

(7.2) $WLxy =_{df} \exists z(Pzy \wedge Lxz)$ (Whole Location)

(7.3) $GLxy =_{df} \exists z \exists w(Pzx \wedge Pwy \wedge Lzw)$.   (Generic Location)

Thus, exact location is a special case of a more general notion of location: within certain obvious limits, if 'L' expresses the notion of a minimal address, 'WL' corresponds to the wider, non-minimal notion (Central Park is wholly located in the region of Manhattan), 'PL' to its dual (Manhattan is partly located at the region of Central Park), and 'GL' to the general case (96th Street is generically located in the region of Central Park). This is captured by the following immediate consequences of (7.1)–(7.3):

(7.4) $Lxy \rightarrow PLxy \wedge WLxy$

(7.5) $PLxy \vee WLxy \rightarrow GLxy$.

With the help of topology we can then complete the picture by introducing further distinctions—for example:

(7.6) $TPLxy =_{df} \exists z(TPzx \wedge Lzy)$ (Tangential PL)

(7.7) $IPLxy =_{df} \exists z(IPzx \wedge Lzy)$ (Internal PL)

(7.8) $TWLxy =_{df} \exists z(TPzy \wedge Lxz)$ (Tangential WL)

(7.9) $IWLxy =_{df} \exists z(IPzy \wedge Lxz)$.   (Internal WL)

A model of all these relations is illustrated in figure 7.1. Of course, in a domain comprising only spatial regions (so that both arguments of 'L' range over entities of the same kind), these locative relations can be made to collapse onto plain mereotopological relations by interpreting 'L' as identity. (Compare figures 3.1 and 4.1.) In that case we get the correspondences

(7.10) $PLxy \leftrightarrow Pyx$

(7.11) $WLxy \leftrightarrow Pxy$

(7.12) $GLxy \leftrightarrow Oxy$,

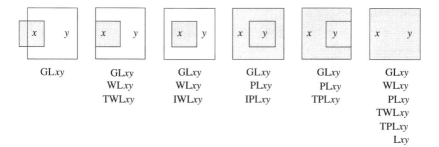

Figure 7.1
The basic patterns of spatial location.

and similarly for the relations defined in (7.6)–(7.9). In general, however, there is more to the ontology of location than just spatial regions, so the relations defined above are all distinct from those of the underlying mereotopological apparatus.

Turning now to the specific axioms for 'L', we begin with two minimal assumptions on the intended interpretation of 'L' as a locative relation:

(L.1) $Lxy \wedge Lxz \rightarrow y=z$ (*Functionality*)

(L.2) $Lxy \rightarrow Lyy.$ (*Conditional Reflexivity*)

(L.1) guarantees that a single entity cannot be exactly located at two distinct regions: L is a functional relation. (L.2) guarantees that L behaves as a reflexive relation whenever it can: all those things at which something is located—i.e., on the intended interpretation, all spatial regions—are located at themselves. This immediately implies that no distinct regions can be exactly co-located:

(7.13) $Lxy \wedge Lzw \wedge Lyw \rightarrow y=w.$

Moreover, (L.1) and (L.2) ensure that L is both antisymmetric and transitive:

(7.14) $Lxy \wedge Lyx \rightarrow x=y$

(7.15) $Lxy \wedge Lyz \rightarrow Lxz.$

Since 'L' is reflexive on all regions, relative to the sub-domain of regions exact location is therefore a well-behaved partial ordering.

Note that we are not assuming that everything is located somewhere, which would be a way of characterizing a world inhabited exclusively by spatial entities. Nor are we assuming that every region is the location of something, i.e., a region at which something is located (besides the region itself). These are substantive principles, to which we shall return. First, we must supplement the pure locative axioms (L.1) and (L.2) by suitable principles bridging the theory of location to its mereotopological background (which we identify with **GEMTC**, the theory of General Extensional Mereotopology with Closure conditions.)

To this end we begin by assuming the following two axioms, which guarantee that the basic mereotopological relations of parthood and connection are properly mirrored in the domain of regions:

(L.3) $Pxy \land Lxz \land Lyw \rightarrow Pzw$

(L.4) $Cxy \land Lxz \land Lyw \rightarrow Czw$.

These axioms imply, for instance, that the parts of an object are wholly located in the object's region, and similarly for tangential and interior parts:

(7.16) $Pxy \land Lyz \rightarrow WLxz$

(7.17) $TPxy \land Lyz \rightarrow TWLxz$

(7.18) $IPxy \land Lyz \rightarrow IWLxz$.

However, (L.3) and (L.4) are still too weak to guarantee a systematic link between the mereotopological properties of a spatial entity and those of its location. For instance, nothing guarantees that an object is partially located in every part of the region at which it is exactly located (the analogue of (7.16) for 'PL'), for nothing guarantees that for each part of the region there is a corresponding part of the object. Likewise, the analogues of (7.17) and (7.18) for 'TPL' and 'IPL' do not follow from (L.3) and (L.4). Accordingly, we assume them as further basic principles:

(L.5) $Pxy \land Lzy \rightarrow PLzx$

(L.6) $TPxy \land Lzy \rightarrow TPLzx$

(L.7) $IPxy \land Lzy \rightarrow IPLzx$.

((L.5) actually follows from (L.6) and (L.7), but we list it here for easier reference.) Together with (L.3) and (L.4), these principles effectively establish the intended meaning of the basic patterns of inexact location. Among other things, they guarantee that all of our derived locative relations satisfy certain intuitive patterns of monotonicity—for instance,

(7.19) $\text{PL}xy \wedge \text{P}zy \rightarrow \text{PL}xz$

(7.20) $\text{IPL}xy \wedge \text{IP}zy \rightarrow \text{IPL}xz$

(7.21) $\text{TPL}xy \wedge \text{TP}zy \rightarrow \text{TPL}xz$

(7.22) $\text{WL}xy \wedge \text{P}zx \rightarrow \text{WL}zy$

(7.23) $\text{TWL}xy \wedge \text{TP}zx \rightarrow \text{TWL}zy$

(7.24) $\text{IWL}xy \wedge \text{IP}zx \rightarrow \text{IWL}zy$.

Other basic properties are easily verified. In particular, inspection shows that 'WL', 'PL', 'IPL', and 'IWL' are all antisymmetric and transitive, like 'L'; 'TPL' and 'TWL' are antisymmetric but not transitive (for instance, Manitoba is TW-located in the region occupied by Canada, which in turn is TW-located in the region occupied by North America, but Manitoba is not TW-located in this last region); and 'GL' is neither transitive nor antisymmetric (the locations of Central Park and 96th Street are generically located at each other, but they do not coincide). The best we can say is that 'GL' is reflexive and symmetric among regions—but this is obvious: with respect to regions, general location is neither more nor less than mereological overlap.

7.3 Regions

The minimal theory defined by (L.1)–(L.7), call it **L**, fixes the intended reading of the locative predicate 'L' against the background of mereotopology. We have also seen that the limited reflexivity of L can be used to pick out a defining property of spatial regions: regions are those things that are located at themselves. Let us, then, be explicit about this and define a 'region' predicate (R) accordingly:

(7.25) $\text{R}x =_{\text{df}} \text{L}xx$.

Since we allow for boundaries—interiorless entities such as surfaces and edges—the predicate defined here is to be interpreted in a most unrestricted way. Its extension is to include voluminous regions (the locations of ordinary voluminous bodies such as chairs and tables) as well as boundary-like regions (the regions corresponding to interiorless entities, such as the surface of a table). Our predicate, therefore, is not to be interpreted in Whiteheadian fashion.[3] We are simply thinking of regions as the places of our spatial entities, the spatial items inhabited by whatever spatial entities we have in the domain of our theory.

Focusing now on this predicate, at least two more structural facts are worth noting. First, (L.2) together with (7.16) ensure that being a spatial region is a dissective property, i.e., the parts of a region are themselves regions:

(7.26) $Rx \wedge Pyx \rightarrow Ry$.

Second, it follows from this that the property of being a spatial region is closed under mereological product, in the sense that the product of two overlapping regions is itself a region:

(7.27) $Rx \wedge Ry \wedge Oxy \rightarrow R(x \times y)$.

This goes some way towards establishing that spatial regions form a mereologically well-behaved domain. However, (L.1)–(L.7) are not sufficient to ensure that a similar fact extends to mereological sums. A domain built up from two atomic regions x and y satisfies the axioms even if $x+y$ is left out of the extension of 'R'. This means that even on the assumption that the left and the right half of a soccer field might be regions, the entire field might not. Hence, to do justice to the intuition that the sum of any two regions is itself a region, we need extend **L** with a further, independent axiom on 'R':

(L.8) $Rx \wedge Ry \rightarrow R(x+y)$.

More generally, (L.8) could be strengthened to cover infinite sums:

(L.8′) $\exists x(\phi x) \wedge \forall x(\phi x \rightarrow Rx) \rightarrow R(\sigma x(\phi x))$.

If there are ϕers, and if all the ϕers are regions, then putting them together is sure to yield a region. In particular, the sum of all regions is itself a region—the universal region. Of course, we are allowing here for the possibility of regions that are not self-connected: if an object is scattered, so will its region. The rationale for (L.8) and (L.8′) is, in fact, related to the fact that the ontological neutrality of the operation of mereological sum cannot prevent regions from being part of hybrid sums, such as John + John's region. This explains why closure under sum is independently necessary for constructing regions out of component regions, even if the dissectivity of the region predicate is automatically granted downwards. (7.26) says that regions are necessary in order to build up regions mereologically; (L.8′) ensures that they are sufficient.

Now, if 'R' is to be interpreted as a characteristic region predicate, much more can be said about its extension. For instance, nothing in the axioms assumed so far implies that spatial regions form a dissective, atomless domain, in the sense that every region has some *proper* part which is a region. If desired, that must also be added as a further independent axiom. More generally, we can express the assumption that space is dissective downwards or upwards by means of the following axioms, respectively:

(L.9a) $Rx \rightarrow \exists y(Ry \wedge PPyx)$ (*Down-dissectivity*)

(L.9b) $Rx \rightarrow \exists y(Ry \wedge PPxy)$.   (*Up-dissectivity*)

An even stronger axiom would be density: the mereological nesting of regions always yields "remainders" that are themselves regions:

(L.10) $Rx \wedge PPyx \rightarrow \exists z(Rz \wedge PPzx \wedge PPyz)$.   (*Density*)

We could also consider, here, an axiom to the effect that *within the domain of regions* the mereological relation of parthood and the topological relation of enclosure coincide (and therefore that 'C' is extensional among regions). This would amount to assuming the R-restriction of the Parthood Axiom (C.10) of Strong Mereotopology (**SMT**):

(L.11) $Rx \wedge Ry \rightarrow (RExy \rightarrow Pxy)$,   (*Region Parthood*)

where

(7.28) $RExy =_{df} \forall z \ (Rz \rightarrow (Czx \rightarrow Czy))$.

We found reasons for rejecting the Parthood Axiom as a general axiom true of all spatial entities; but nothing said so far goes against its restriction to regions. In fact we think (L.11) is most reasonable—but with two provisos. First, (L.11) is only acceptable on the assumption that x is not an atom or a boundary: otherwise x would be part of its complementary region, which is impossible.[4] Second, our standard interpretation of the connection relation as holding between two entities just in case one overlaps the closure of the other (or vice versa) might falsify (L.11), unless we assume that all regions are topologically "regular." Even if we assume that no region may possess outgrowing boundary hairs and may not consist of voluminous parts connected by a boundary, as suggested in relation to the idea that boundaries are ontologically dependent entities (section 5.8), we can still falsify (L.11) by taking x to be a closed region and y any region consisting of all parts of x minus some interior isolated point. To account for these two provisos, then, let us define a regularity predicate along the lines of ordinary topology,

(7.29) $\mathrm{R}gx =_{\mathrm{df}} \mathrm{P}(cx, c(ix)) \wedge \mathrm{P}(i(cx), ix)$,

and let us introduce a predicate singling out those regions that are both regular and extended, i.e., not atomic or boundary-like:

(7.30) $\mathrm{ERR}x =_{\mathrm{df}} \mathrm{R}x \wedge \mathrm{R}gx \wedge \exists z \mathrm{IPP}zx$.

Then the proper form of the Parthood Axiom for regions is

(L.11′) $\mathrm{ERR}x \wedge \mathrm{ERR}y \rightarrow (\mathrm{RE}xy \rightarrow \mathrm{P}xy)$.

Other axioms may be added at this point to capture further characteristic properties of regions and, more generally, of the mereotopology of space. However, there is no need here to pursue this line further. Let us see instead how the structure of regions is related to that of their tenants.

7.4 The Regions of Things

Let us assume that every entity is located at some region. As was mentioned in section 7.2, this has the drastic effect of excluding all entities lacking spatial location (such as sets and numbers) from the domain of our theorizing. But here we are only interested in the structure of spatial location, so such a restriction is methodologically convenient:

(L.12) $\exists y \, \mathrm{L}xy$.

Given any entity x, we can then speak of its region rx, the region where x is exactly located:

(7.31) $rx =_{\mathrm{df}} \iota y \, \mathrm{L}xy$.

The unicity of rx follows directly from the Functionality Axiom (L.1). (Thus, given the immediate consequence

(7.32) L$xy \leftrightarrow y=$rx,

in the presence of (L.12) one could also take 'r' as primitive and define 'L' accordingly.) From (L.2), together with (L.3)–(L.7), it follows further that r is idempotent and distributes over sums:

(7.33) r(rx) = rx

(7.34) r($x+y$) = r$x+$ry,

whence

(7.35) r($x+y$) = r($x+$ry) = r(r$x+y$) = r(r$x+$ry).

That is, objects and their regions do not pile up to form other regions. By contrast, the analogues for the operation of mereological product may not hold unless the domain is restricted to include only regions. John and r(John) have no parts in common (or so we may assume). In fact, if we confine ourselves to the domain of regions, besides idempotence and distributivity we also have expansiveness (trivially):

(7.36) R$x \rightarrow$ Px(rx).

In view of (7.33)–(7.35), this means that in the domain of regions the r operator behaves as a closure operator (see (C.5)–(C.7)). However, this is not very interesting in itself, since (7.36) can obviously be strengthened to

(7.37) R$x \rightarrow x=$rx,

which marks a complete collapse of the distinction between an entity and its region. Not a surprise: the notion of an object's region is of interest only when the object is not itself a region.

Now, how is the mereotopology of spatial objects mirrored in that of their regions? Observe that the basic postulates of **L** immediately imply the following:

(7.38) P$xy \rightarrow$ P(rx)(ry)

(7.39) C$xy \rightarrow$ C(rx)(ry).

In fact, a similar result applies to many other relations besides the primitives 'P' and 'C', indeed to all positive mereotopological predicates such as 'O', 'U', 'IP', and 'TP'. These implications are trivial if x and y are both regions, but when it comes to other spatial entities this structural mirroring is crucial. In particular, the converses of (7.38), (7.39), and the like need not hold when x and y are not regions, and the whole point of a theory of location is precisely to do justice to the idea that the mereotopology of spatial entities is mirrored by, but does not reduce to, the mereotopology of their regions. This was the gist of our conclusion in chapter 2 and we are now finally in a position to take a look at the general picture. First, however, let us note that principles such as (7.38) and (7.39) are not enough to guarantee a full agreement between mereotopological and locative structures. To this end we also need to guarantee that the topological operators of boundary, closure, and interior are also well-behaved. For instance, it seems reasonable to expect that the region of an entity's boundary is the boundary of that entity's region. As it turns out, however, neither the minimal theory **L** nor the extensions of **L** considered so far are strong enough to guarantee such a correspondence. If we wish to secure it, we must therefore assume it as an independent axiom:

(L.13) $r(b(x)) = b(r(x))$.

Corresponding facts about the closure and interior operators will then follow using the full strength of **GEMTC**:

(7.40) $r(c(x)) = c(r(x))$

(7.41) $r(i(x)) = i(r(x))$.

7.5   Location, Connection, and Parthood

Let us pause now and see what we have got. We have helped ourselves with three basic relations: parthood (for the mereological aspects of spatial representation), connection (for the topological aspects), and location. Some basic structural relationships among these three notions have now been examined. But the general question of their mutual interdependence is still open. Is there any mereological or topological relationship between an entity and the region where it is located? Is any of these three relations dispensable in term of the others? In the domain of regions, location reduces to mereology by (7.37). And if we only focus on regular extended regions, mereology in turn reduces to topology by (L.11′), so only the latter is actually needed. These are in themselves significant facts. But what about the general case?

Consider connection. We have been working with a standard interpretation of 'C', whereby two things are connected when they are either overlapping or abutting, i.e., when at least one overlaps the closure of the other. One might be tempted here to go beyond this interpretation and to rely on a more general notion of topological connection. If x is located at rx—one could argue—this would be *because* x is connected with rx. However, what should the relevant notion of connection between x and rx amount to? Overlap is excluded: an object does not share parts with the region at which it is located—unless of course the object is itself a region, or a mereological sum including regions among other things. Moreover, two objects can share at least part of the same location without sharing any parts. Think again of the stone inside the hole: the stone is not part of the hole, even though the region of the stone is part of the region of the hole. Abutting is also excluded, if this is understood as the relation of *external* connection holding between contiguous entities. To fix ideas, we must therefore think of the relevant relation as a sort of *internal* connection. But this would take us into a circle. For what kind of relation could that be, if not a relation of location?

We face here an apparent limit of topology as a general theory of space. Topology—and, more generally, mereotopology—falls short of expressing one fundamental metaphysical fact about space, namely embedding in space. The *analysis situs* overlooks the fact that objects are *situated*, and this is why a theory of location seems to be independently needed. This is not the whole story, though. For consider again (7.39): if two objects are connected, so are their regions. Why not assume the converse as well? Why not regard connection between two regions as a sufficient condition for connection between the regions' tenants? This would mark an extension of the standard interpretation of 'C'. But it would be definite enough, and it would strengthen the tie between location and topology: If we read 'Cxy' as 'the region of x and the region of y are either overlapping or abutting', then location is connection of a kind—or so one could argue. It is connection minus overlap minus abutting. For consider again x and its region, rx. On the usual interpretation of 'C' they are not connected: they do not share any boundaries (we may suppose). On the new interpretation they would be connected. Now, we just saw that x and rx do not overlap and we do not want to say that they are externally connected either. This would indeed follow from the definitions of chapter 4, but if we strengthen (7.39) to a biconditional, the definition of 'EC' must obviously be emended:

(7.42) $ECxy =_{df} Cxy \wedge \neg O(rx)(ry)$.

So again x and rx would be connected, non-overlapping, and non-abutting. But now we would have an account of each of these notions. *Ergo*, we would after all have an account of the locative relation between x and its region—a purely mereotopological account.

Well, what really follows is that location can be explained in terms of 'C', 'P', *and* 'r'. But we already knew that regions are essentially linked to location. The region of x is the region at which x is located (7.31), and x is located at its region and nothing else (7.32). So, assuming 'r' is tantamount to assuming 'L' and the limits of mereotopology show up again. A location primitive must be assumed in some form or other.

On the other hand, consider parthood. It is clear that on this extended interpretation of 'C', mereology and topology would also retain an independent status. For C would be a proper extension of O by assumption. And conversely, with (7.39) strenghtened to a biconditional, everything connected to an entity x would be connected to its region rx, and yet the former is not part of the latter, so E (topological enclosure) would be a proper extension of P. Likewise, if a stone is inside a hole, then everything connected to the stone would be connected to the hole even if the stone is not part of the hole. Now let us go back to the standard interpretation of 'C'. If connection is really only a matter of overlap or abutting, then the stone and the hole are arguably disconnected even though they are partly co-located. They share no boundaries and bear no mereotopological relations even though they literally interpenetrate each other. But what about the stone and its region—are they connected? Is one topologically enclosed in the other (i.e., is everything connected to the stone connected to its region)? And what about other cases of (partial) co-location, for instance, what about the rotating and the cooling down of the Earth—are they connected? Assuming that these are two distinct events, do they share any boundaries with their common region? If the answer is negative in all cases, then an important consequence follows.

For then the reducibility of parthood to connection acquires a certain plausibility. If there are no pairs of entities x and y such that x is not part of y even though everything connected with x is connected with y, then the Parthood Axiom (C.10) becomes attractive. The restriction to regions could be dropped from (L.11′) and the reduction of mereology to topology would go all the way, *at least as far as regular voluminous bodies are concerned*. If, on the other hand, we leave open the possibility that some such cases of co-location involve boundary-sharing (if not sharing of parts proper), then there is no reduction and the final picture is truly threefold: we need location, we need topology, and we need mereology. Which of these accounts is correct is perhaps a matter of theoretical stipulation. We prefer the open choice, as we argued earlier; hence the threefold picture.

7.6   Sharing an Address

So far we have been dealing with location understood as a relation between an object and a region of space. But we can also refer to the notions introduced above to express the fundamental ways in which two objects can be related *to each other* in terms of location. For every locative relation $\mathcal{R}$ (among GL, WL, PL, IWL, IPL, TWL, TPL), we can define the corresponding relation of region-location:

(7.43) $R\mathcal{R}xy =_{df} \mathcal{R}x(ry)$.

As the following equivalences show, any relation thus defined is equivalent to a relation expressible via P, C, and r:

(7.44)   RPL$xy \leftrightarrow$ P(ry)(rx)

(7.45)   RWL$xy \leftrightarrow$ P(rx)(ry)

(7.46)   RGL$xy \leftrightarrow$ O(rx)(ry)

(7.47)   RTPL$xy \leftrightarrow$ TP(ry)(rx)

(7.48)   RIPL$xy \leftrightarrow$ IP(ry)(rx)

(7.49)   RTWL$xy \leftrightarrow$ TP(rx)(ry)

(7.50)   RIWL$xy \leftrightarrow$ IP(rx)(ry).

But on the face of it, (7.43) represents a fundamental key for extending the theory of location developed so far. For it makes it possible to relax the constraint that location is essentially a relation between objects and regions. By (7.43), we can speak of an object being wholly, partly, or generically located at/within another object, and likewise for the other relations. (We can say, for instance, that John is internally wholly region-located (RIWL) in his godfather's house, or that a stone at the bottom of a hole is tangentially wholly region-located (RTWL) in the hole.)

One more relation that can be defined along these lines is *exact co-location*, which so far we have been using intuitively:

(7.51) $RLxy =_{df} Lx(ry)$.

The link with (7.44)–(7.50) is given by the fact that exact co-location amounts to identity of corresponding regions:

(7.52) $RLxy \leftrightarrow rx = ry$.

Thus, RL is an equivalence relation (reflexive, symmetric, and transitive). Moreover, each relation $R\mathcal{R}$ in (7.44)–(7.50) is linked to RL in a way parallel to the definition of $\mathcal{R}$ in terms of L:

(7.53) $RWLxy \leftrightarrow \exists z(Pzy \land RLxz)$

(7.54) $RPLxy \leftrightarrow \exists z(Pzx \land RLzy)$

and so on. (Compare (7.1)–(7.3) and (7.6)–(7.9).) And each relation $R\mathcal{R}$ satisfies the analogue of the basic properties of $\mathcal{R}$:

(7.55) $Pxy \land RLyz \rightarrow RWLxz$

(7.56) $Pxy \land RLzy \rightarrow RPLzx$

and so on. (Compare (7.16)–(7.18) and (L.5)–(L.7).) This is as it should be. If the stone exactly fills the hole, and is therefore exactly co-located with it, then any part of the stone is wholly located within the (region of the) hole, and the hole is partly located at the (region of the) stone's part. Moreover, by (7.34) the sum consisting of the stone *and* the hole is exactly co-located with the stone, and also with the hole. More generally:

(7.57) $RLxy \rightarrow RLx(x+y)$.

This clarifies another important sense in which mereotopological structures are mirrored in locative structures: if the sum of two things is nothing over and above the two things, then its behavior with respect to location cannot run afoul of the behavior of its constituent objects. If the stone and the hole are located in the same region, then so is their mereological sum.

Another important aspect of (7.57) is that something may be co-located with some of its proper parts. It is here that co-location falls short of identity. In this respect co-location is rather like the relation of coincidence mentioned in connection with Brentano's account of boundaries: boundaries are *located in* space but do not *occupy* space, and can therefore coincide; they can be perfectly co-located with one another.[5] There is an important difference, however. Boundaries do not occupy space insofar as they do not *take up* space—they are lower-dimensional spatial entities. Here, in contrast, we are interested in a general feature of spatial location, namely, that more entities can share the same address, and this feature applies to all sorts of entities regardless of their dimensionality. Every spatial entity has an address; but to have an address does not mean to be the exclusive *owner* of it.

7.7 Owning an Address

When can something be said to *own* its address? Intuitively, something owns its address if it does not share it with other things. Thus material objects—such as cars in a parking lot—seem to have the property of excluding other material objects from the region they are located at. For them, location is exclusive—they *occupy* the regions where they are located. And this fact has important bearings as to matters of identity. Provided it never moved, the car that Mary parked in the space numbered 21 is the same as the car that she is now getting in. This was the gist of Locke's identity principle, discussed in section 2.2.

To own an address, then, is to occupy it, and *occupation* is a form of exclusive location. As a tentative step, we may consider characterizing it thus:

$$(7.58) \quad \text{OC}xy =_{df} Lxy \wedge \forall z(Lzy \rightarrow z{=}x),$$

or, using the notion of co-location (RL):

$$(7.58') \quad \text{OC}xy =_{df} Lxy \wedge \forall z(\text{RL}zx \rightarrow z{=}x).$$

This characterization is defective in several ways, however. For one thing, we know that L is reflexive on spatial regions (and RL refexive *simpliciter*), so in general we would have to insert in the first conjunct of these definitions the proviso that $x{\neq}y$, and in the antecedent of the second conjunct the proviso that $z{\neq}y$. Secondly, the consequent of the second conjunct must be strengthened to an overlap proviso, i.e., 'Ozx'. Otherwise OC would be empty. (Suppose x is a stone that exactly fills a hole z: x and z are co-located and distinct, and yet we want to say that x truly occupies $r(x)$.) Thirdly, and more generally, (7.58) and (7.58′) fail to fix a suitable range for the quantified variable 'z'. To draw again on Locke, no two things *of the same kind* can be located in the same place at the same time; but a hole and its filling stone may very well be exactly coincident in space, at least for some portions of their respective lives. Hence, exclusive location (dismissing only regions of space) is too strict a condition for any interesting notion of occupation—the notion whereby material objects do, and holes do not, occupy the space at which they are located. After all, if the hole can be interpenetrated by the stone, the stone can be compenetrated by the hole. The difference is that the hole can be compenetrated by other holes (as when you put a wedding ring inside the hole of a doughnut); but the stone cannot be compenetrated by other stones, or by other material objects at large. What we need is thus a relative notion, one that captures the sortal nature of Locke's criterion. And the natural solution is to use a definition schema. Where ϕ is any condition, we may say that an object ϕ-occupies a certain region if and only if it is the only ϕer that is located at that region. We thus arrive at

(7.58″) $OC_\phi xy =_{df} \phi x \wedge Lxy \wedge \forall z(\phi z \wedge Lzy \rightarrow Ozx)$,

which immediately implies

(7.59) $OC_\phi xy \wedge OC_\phi zy \rightarrow z{=}x$.

Together, (7.58″) and (7.59) provide the desired characterization of the distinguishing feature of occupation *versus* mere location.

Of course, being defined in terms of exact location, spatial occupancy is in fact *exact* spatial occupancy. A chair occupies exactly the region of space at which it is exactly located. Weaker notions of occupancy can then be defined matching the weaker notions of location defined in (7.1)–(7.3) and (7.6)–(7.9), but we need not go into these routine details. Suffice it to remark that there is a slight asymmetry in this regard, as location and occupation have opposite part-whole structures. Something x is partly located at a region y if some part of x is located at y; but if this part actually occupies y (relative to some relevant condition ϕ), we would rather say that x *wholly* occupies y; by contrast, x *partly* occupies y if x occupies only part of y, in which case we say that x is wholly located at y. Thus, the correct analogues of (7.1) and (7.2) for occupation are

(7.60) $POC_\phi xy =_{df} \exists z(Pzy \wedge OC_\phi xz)$

(7.61) $WOC_\phi xy =_{df} \exists z(Pzx \wedge OC_\phi zy)$,

which in turn imply

(7.62) $POC_\phi xy \rightarrow WLxy$

(7.63) $WOC_\phi xy \rightarrow PLxy$.

There is a deeper asymmetry between occupation and location in this connection. Occupation is first and foremost exact occupation: to understand what it is for something to occupy a region requires in an essential way the capacity to compare the respective shapes and dimensions of the occupying thing and of the occupied region. By contrast, to understand what it is for a certain thing to be located at a certain region requires the competence for somebody to find a place (a temporary address) for that thing, but we have seen at the beginning that this leaves room for some flexibility as to the actual extension of the relevant place. We can distinguish a minimal and a non-minimal address, with all the intermediate degrees allowed by this contraposition. The notions of whole, partial, generic, but also tangent and internal location are meant to capture the variety of ensuing plausible relations; yet some of the corresponding notions of occupation would hardly enjoy any plausibility, except from a purely algebraic perspective. What would it be for something to internally occupy a region of space? And to generically occupy one?

With these provisos, two last facts about the notion of occupation characterized by (7.58″) are worth noting. First, if ϕ is the property of being a region (R), then every region is actually such as to ϕ-occupy itself (by (L.1)):

(7.64) $OC_R xx$.

This may be regarded as a case of *sui generis* occupation, for spatial regions can hardly be said to *occupy* anything. But the connection with material objects is not uninteresting: objects cannot share an address with other objects; regions cannot be located at *other* regions. This is also expressed by

(7.65) $OC_R xy \rightarrow x = y$,

whence we immediately infer that OC_R is both symmetric and anti-symmetric, and that it collapses to identity.

The second fact concerns (7.59), the principle that two distinct ϕers cannot occupy the same region. We may follow Locke and regard this as a sort of identity principle apt for the characterization of material objects. That is, we may take it that when ϕ is the property of being a material object, the predicate 'OC_ϕ' is not vacuous and (7.59) expresses a substantive truth. Yet this would not be a truth of the theory of location. Nothing in the theory rules out the possibility that two—or, for that matter, infinitely many—distinct indistinguishable material objects be perfectly co-located. If occupation is exclusive location, relative to things of a kind, then material objects are illustrious candidates to the role of spatial occupants. But it is not the task of a theory of location to determine *whether* such entities are actually such as to occupy space. Nor should the theory be concerned with the task of explaining *why* material objects occupy space, if they do, and why they are therefore impenetrable by other material objects. These are metaphysical questions, and in their regard the theory of location is uncommitted.[6]

8 Empty Places

Location, as we have characterized it, is a relation between a spatial entity and its region. But location is also, in an important sense, a relation linking directly one spatial entity to another. Sometimes this reduces to the first sense of location: the stone is located in the hole because it is located (wholly) in the region of the hole; John's foot is located in his body because the foot is part of the body, and is therefore located (wholly) in the region where the body is located. This was the point of the region-location relations introduced in section 7.6. But we may also want to speak of an object being located in another without there being any corresponding relationship between their regions, and there seems to be no straightforward way of dealing with this notion in terms of the apparatus of chapter 7. The fly is in the glass, but there is no relation between the region of the fly and that of the glass. And a worm can be inside a piece of cheese without there being any relation between the region of the worm and that of the cheese: the worm lies *outside* the cheese, strictly speaking, and yet it is located *inside* it. Indeed, this is generally to be expected in view of the constraints on spatial occupancy: if x and y are the sort of entities that occupy their spatial regions, then the regions they occupy must be discrete unless the objects themselves overlap mereologically.

In this chapter we consider a way of dealing with this second, extended notion of location. The main key—we argue—is again the interpenetrability of certain kinds of entity: in many cases, to be located *in* an object is to be located (wholly or partly) in a hole of that object. And to be located in an object's hole is to be located *outside* the object, at a region disjoint from the object's region. This will yield an account that is still very close to the basic apparatus developed in the previous chapter. And it will also allow us to give one more concrete illustration of the advantages of a pluralistic spatial ontology. Taking holes seriously makes it possible to disclose conceptual distinctions that would be left in the dark if one focused only on paradigmatic cases such as spatial regions or material objects. By realizing that spatial entities might be of many different kinds (material and immaterial, in this case), one realizes that they might also bear many different relations to space. And this in turn helps sharpen the main coordinates in the structure of spatial representation.

8.1 The Geometry of Containment

Consider the fly in the glass. If all we can work with are the fly, the glass, and the regions that they occupy, the options are rather slim. Between the region of the fly and that of the glass there may subsist no mereotopological relation of the sort described so far.

One possibility would be to reason in terms of idealized convex hulls.[1] The convex hull of an entity x is, intuitively, that entity that would be encompassed if x were wrapped in a tight rubber membrane. And one might suggest that the relevant locative relation can be explained in terms of mereological inclusion in the relevant convex hull: something x is located in something y if the region of x is part of the region of the convex hull of y (intuitively: if x would be trapped inside the membrane around y). This certainly captures a basic intuition about containment. With reference to figure 8.1a, the fly is in the glass because its region is part of that of the convex hull of the glass; not so for the ant, which is then *not* in the glass.

Unfortunately, as Herskovits (1986) points out, this approach fails to appreciate the essential role of containing parts as opposed to other non-convex parts. Think of a fly near the stem of a champagne glass: it may well fall within the convex hull of the glass, but that does not make it a fly *in* the glass (figure 8.1b). In other words, reference to the convex hull might be necessary but not sufficient to single out those regions that are relevant for the purpose of spatial inclusion. Focusing exclusively on the convex hull of the object's *containing* parts (as considered, e.g., in Vandeloise 1986) is likewise inadequate. It is not the seeming circularity that undermines this suggestion: we may well assume that we have an independently clear notion of what counts as a containing part. The trouble is that containing parts come in many shapes. The outer surface of a containing part may itself involve concavities. And as Vieu (1991: 207) points out, this is enough to give rise to counterexamples. (See figure 8.1c.)

Figure 8.1
Inclusion in the convex hull (dashed line) is not an adequate criterion for being in the glass.

Now, from our present perspective the difficulty that we are facing is halfway between a general limitation of mereotopology, on the one hand, and a fundamental problem that we already saw in our brief discussion of Hoffman and Richards' theory of negative parts (section 2.5), on the other. It is a problem of general theoretical limitations inasmuch as the relevant role of what really counts as a container cannot be explained in mereotopological terms, even if we rely on an auxiliary convex hull operator. A container is a fillable thing, and fillability is a complex dispositional property that cannot be explained in terms of simple part-whole structures. And the problem is also similar to the dilemma of the negative-part theory insofar as it requires thinking about *distinguished* negative parts. On the negative part theory we could identify the convex hull of an object with the sum of its positive and negative parts. But not just any negative part can serve the purpose of containing an object, unless of course we identify the missing parts with the containing parts. Why is a fly near the stem of a champagne glass not in a negative part of the glass?

It is here that holes enter the picture. By a hole we mean, in this context, any voluminous chunk of an object's complement that we would ordinarily call a hole, but also, much more broadly, any part of the complement determined by a fillable discontinuity in the object's surface. Let us, indeed, identify holes precisely with those parts of an object's complement that are encompassed by the object's containing parts. What counts as a containing part is not at issue here, as we said: we assume that to be an independently clear notion. (It certainly is not a notion that can be explained purely in mereotopological terms.) The point is, rather, that the geometry of containment can be explained, not by reference to the containing part, but by reference to the associated hole. The fly is located in the glass if and only if it occupies a regions that is part of (or at least overlaps significantly) the region of the hole in the glass. This is why in cases *b* and *c* of figure 8.1 the fly is not in the glass: it is located within a region included in the convex hull of the relevant containing part; but not every such region is the region of a hole; and only those regions that are must be taken into account to represent spatial containment.

In making this suggestion, we take it of course that the containing part of the glass is the part where you pour the wine. *If* we took the space around the stem as a containing part, then the fly *would* be in the glass, on this as on any reasonable account. However, it bears emphasis that our suggestion is not meant to imply that every case of hole filling is a case of spatial containment, flexible as the relevant notion of a hole may be. If the glass has a handle, the hole defined by the handle should not count for the purpose of deciding whether the fly is in the glass. (See figure 8.2.) Perhaps the stem or the containing part itself is perforated: then, again, a fly in such a hole (perforation) would not be in the glass. Or again, to use a related example from Vandeloise 1994, the bulb is in its socket, but the bottle is not in its cap—even though the geometry of the socket is identical to that of the cap, and the geometry of the bulb is identical to that of the bottle. (In other cases, filling a hole is not even a necessary condition for containment: the flowers are in John's hands, but there is no hole involved here unless we stipulate otherwise.) It is apparent that these cases reveal the limits of the approach insofar as it is purely geometric: a full account calls for a step into other territories where pragmatics, or functional and causal factors at large, must be taken into account. Our

Figure 8.2
Functional considerations are involved in selecting the relevant holes: in each case here the fly is in a hole in the glass, but not in the glass itself.

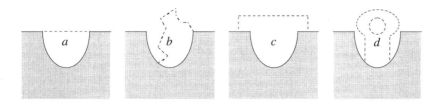

Figure 8.3
Only *a* is a hole; *b*, *c*, *d*, and the like violate the intuitive condition that a hole's fiat boundary
is determined by the convex hull of the object.

point is that explicit reference to holes can mark an improvement as far
as the *geometric part* of the story goes. True, only some holes count for the
purpose of representing containment. But which holes do count is not a
question for the geometric analysis of the problem.

Let us also clarify a link between the hole concept and the notion of
convex hull. To be in a hole is to be region-located within the boundary
of the hole. But what exactly is the boundary of a hole? As was seen in
section 5.7, part of the boundary is supplied by the material object at
whose surface the hole is located: the inner surface of the containing part
bounds the hole from outside. (So the hole is open there.) But with the
exception of holes that are completely hidden inside an object, such as a
bubble inside a wheel of Swiss cheese, the boundary of a hole typically
includes also an immaterial part: the fiat boundary that separates it from
the rest of the object's complement. Where exactly is *this* boundary
located? Broadly speaking, our answer is that this boundary is determined by the boundary of the convex hull of the relevant containing part.
It corresponds to the minimal surface determined by the rim of the
hole—to the boundary of the ideal perfect filler of the hole.[2] This is
stipulative to some degree. But it helps sharpening the intended interpretation of the basic concepts at issue. There are no holes corresponding to
the bizarre shapes of *b–d* in figure 8.3.

8.2 Holesome Relations

On this basis—and within these limits—let us give a more precise formulation of the account. Let us write 'H*xy*' for '*x* is a hole in *y*' (*y* may be
called the 'host' of *x*). Then the new relation of location ('IN') may be

defined as that relation that holds between two entities when one is region-colocated with a hole in the other:

(8.1) $INxy =_{df} \exists z (Hzy \wedge RLxz)$.

This is meant to reflect the notion of *exact* location (perfect fit). Then we define, more generally,

(8.2) $\mathcal{R}INxy =_{df} \exists z (Hzy \wedge R\mathcal{R}Lxz)$,

where '$\mathcal{R}$' is any locative modifier of the sort familiar from chapter 7 (so that $\mathcal{R}L$ is any relation among GL, WL, PL, IWL, IPL, TWL, TPL, and possibly other relations as well). To illustrate, in figure 8.1a the fly is *wholly* located in the glass: it stands with the glass in the relation WIN. A pencil, by contrast, will only be *generically* located in the glass (GIN). Note that since everything is region-colocated with itself, (8.1) implies that every hole is exactly located in its host. This is all right, but it may be useful to distinguish between proper and improper location, as it were. Object x is properly $\mathcal{R}$-located in object y if it is $\mathcal{R}$-located in y but is not itself a (part of) a hole in y:

(8.3) $P\mathcal{R}INxy =_{df} \mathcal{R}INxy \wedge \neg\exists z (Hzy \wedge Pxz)$.

We do not, here, provide a full axiomatic treatment of 'H' and of its interaction with mereological and topological concepts.[3] Let us simply point out that 'H' will have to obey certain basic principles to ensure a proper relationship of holes with their hosts. For instance, we certainly want to assume that the host of a hole is not itself a hole:

(H.1) $Hxy \rightarrow \neg Hyz$.

Among other things, this guarantees that H is an irreflexive and asymmetric relation; no hole is a hole *in itself*, and there is no room for hole-host loops:

(8.4) $\neg Hxx$

(8.5) $Hxy \rightarrow \neg Hyx$.

Also, there is no mereological overlap between a hole and its host; the hole is part of the host's complement, though a part that must perforce be connected to the host:

(H.2) $Hxy \rightarrow ECxy$.

And there is no overlap between the regions of the hole and the region of the host—these too are externally connected:

(H.3) $Hxy \rightarrow EC(rx)(ry)$.

(Holes can be interpenetrated, but they cannot be interpenetrated by their own hosts.) This ensures that no $\mathcal{R}IN$ relation is reflexive: nothing is located at itself in the sense here at issue.

Further locative relations can at this point be defined. To this end, let us introduce the auxiliary notion of a "containing hull"—the natural counterpart of a convex hull. This can be defined in terms of an operator k yielding the fusion of any given object with all of its holes:

(8.6) $kx =_{df} \sigma z\ (Pzx \lor Hzx)$.

(If holes are always wholly located inside the convex hull of their hosts, the containing hull of an object is always wholly located inside its convex hull, but the converse may fail.) Let us also expand the set of locative predicates of chapter 7 by adding a predicate for the relation of *external region location*—intuitively, region location at a boundary:

(8.7) $RELxy =_{df} EC(rx)(ry)$.

Then we can define, for instance:

(8.8) $JINxy =_{df} WINxy \land RELx(\sim ky)$ (Just inside)

(8.9) $JOUTxy =_{df} RELx(ky) \land \neg RELxy$ (Just outside)

(8.10) $WOUTxy =_{df} \neg RGLx(ky)$ (Wholly outside)

(8.11) $POUTxy =_{df} GINxy \land \neg WINxy$.   (Partly outside)

The exact interpretation of these predicates depends on the issue of the belongingness of the boundary of a hole. (See section 5.7.) If the fiat boundary that separates the hole from the rest of the host's complement is assigned to the complement (i.e., if the hole is open), then no closed objects can be just inside a container. If, by contrast, we assign the boundary to the hole, the situation is reversed and no closed objects can be just outside. Assuming the former interpretation, the natural transition from the two possible extreme positions (corresponding to the relations WOUT and WIN), is illustrated in figure 8.4.

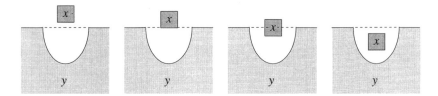

Figure 8.4
Natural transition of an object x from wholly outside (WOUT) to wholly inside (WIN) an
object y. The intermediate steps correspond to JOUT and POUT (in the order).

8.3 Modes of Containment

More relations can be defined by fully exploiting the algebra of spatial
location, as detailed in chapter 7. This will support rich taxonomies such
as the ones defined in Cohn, Randell, and Cui 1995 using the convex-hull
approach mentioned above. In addition, several useful refinements can
be introduced, including some applications to naive-physical reasoning
about containment (in the spirit of Hayes 1985b).

To this end, let us observe that mereotopology allows us to express
certain fundamental differences between holes of different kind. There
are three main kinds (figure 8.5): internal holes (cavities) which are
completely hidden inside the host, and which therefore mark a splitting
in the host's complement;[4] perforating holes (tunnels) which introduce
non-eliminable topological discontinuities (determining an increase in
the topological genus of the host); and superficial holes (or hollows)
which correspond to simple depressions or indentations in the surface of
the host, and could in principle be eliminated by elastic deformation.
These distinctions can be uniformly expressed in terms of the intuitive
notion of an "entry boundary."[5] Let an entry boundary (EB) of a hole x,
relative to a given host y, be a maximally self-connected free boundary

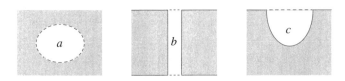

Figure 8.5
Internal hole, or cavity (a); perforating hole, or tunnel (b); superficial hole, or hollow (c).

(FB) of the hole, i.e., a maximally connected part of the hole's (*fiat*) boundary that is nowhere a boundary of the host:

(8.12) $\text{FB}zxy =_{df} Bzx \wedge \neg Bzy \wedge SCz$

(8.13) $\text{EB}zxy =_{df} \text{FB}zxy \wedge \forall w(\text{FB}wxy \wedge Cwz \rightarrow Pwz).$

(We shall only have use for these definitions when x and y are as specified.) Then cavities are those holes that have no entry points: the boundary of a cavity is entirely part of the host's boundary, and there is no way you can get out of the cavity without digging through the host. Hollows, on the other hand, do have an entry point: you can easily pour water into them. In fact hollows have exactly *one* entry point, and this distinguishes them from perforations, which always involve at least two entry points. (If you pour water into them, it can run out from the other side.) Formally, then, the basic distinction between internal holes (IH), perforating holes (PH), and superficial holes (SH) may be characterized as follows:

(8.14) $\text{IH}xy =_{df} Hxy \wedge \neg \exists z \text{EB}zxy$

(8.15) $\text{PH}xy =_{df} Hxy \wedge \exists z \exists w(\text{EB}zxy \wedge \text{EB}wxy \wedge \neg Czw)$

(8.16) $\text{SH}xy =_{df} Hxy \wedge \neg \text{IH}xy \wedge \neg \text{PH}xy.$

We can now exploit these distinctions to introduce corresponding distinctions among locative relations. For instance, we can account for the difference between loose inside and strict, *topological* inside, the latter occurring only in the presence of internal holes. The generating schema is

(8.17) $\text{T}\mathcal{R}\text{IN}xy =_{df} \exists z(\text{IH}zy \wedge R\mathcal{R}\text{L}xz),$

where $\mathcal{R}$ is as in (8.2). Of course this implies that topological inside can never be partial unless the relevant object (the "guest" of the hole) is scattered, and that something is wholly topologically inside an object only if it cannot move to the outside without cutting through the object itself (think of a maggot eating its way out of a cavity in a wheel of Swiss cheese).

In a similar manner we can account for other cases in which the guest object cannot be let free without cutting. Topological inside defines one such case (figure 8.6*a*); another is what may be labelled 'locked inside' (*b*): the hole is open and the object inside it is or can be put in contact

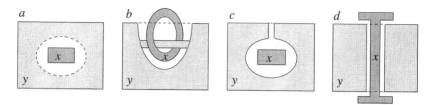

Figure 8.6
Patterns of constrained inside. In some cases the constraint is purely topological (a, b); other cases depend crucially on the morphology of the contained and containing objects (c, d).

with the outside, but the presence of a topological discontinuity constrains its freedom to move.

(8.18) $\text{LIN}xy =_{\text{df}} \exists z(\text{H}zy \land \text{RGL}xz \land \forall w(\text{RL}wz \rightarrow$
$\exists u(\text{PH}uw \land \text{RIWL}u(ky) \land \text{RPL}u(kx))))$.

From a naive-physical perspective this is perhaps one of the most important patterns of interaction concerning holes: keeping material objects (fillers) in place, or hindering their movement. There are of course many other such patterns besides topological constraint. For instance, a plugger is typically a complete, non-exact filler some parts of which are externally connected with the host but not with the hole, so that its translational freedom (Shoham 1984) is constrained. Or take cases c and d in figure 8.6. Here the guest object is kept inside the host by virtue of morphological constraints. Unlike a and b, the guest could be let free without altering the topology; but unlike the basic pattern in figure 8.4, x can be let free only if y or x itself is suitably deformed. Again, we reach here a point where pure mereotopological reasoning shows its limits—shapes become crucial.

8.4   More on Negative Parts

The treatment of locative relations in terms of holes is a good illustration of how representational and ontological aspects may interact. As we said, much depends on how 'H' is cashed out, hence, ultimately, on how holes are construed. Our guiding idea has been that holes are immaterial bodies which can be interpenetrated by other objects. And clearly holes must be immaterial for the account to be viable at all: if holes were construed as material bodies, for instance by identifying them with hole-linings (as in Lewis and Lewis 1970), nothing could be in an object by being *inside* a hole.

Now, we have also insisted that the immateriality of a hole goes hand in hand with its being part of its host's complement: the hole is externally connected to the host (H.2). In fact this is why we had a problem in the first place: the lack of any mereological relation between the region of the fly and that of the glass is not negotiable. Our reasons for this were that treating holes as distinguished parts of their hosts, as in Hoffman and Richards's (1985) theory, would require some criterion for identifying the relevant negative parts. But now one could object that our predicate 'H' suffers from a similar problem. Conversely, one could say that 'H' could be interpreted as a relation of negative parthood, and that therefore one could reduce the sort of locative relation considered above to the locative relations discussed at length in chapter 7. The fly would be region-located in the glass after all. Let us then have a second look at the idea of a negative part.

Consider for a moment how the mereological module operates on holes construed as immaterial bodies of their own, as suggested here. Anytime there is a hole in an object there is some mereological composition around. ('H' is a binary predicate, and the second argument cannot take holes as values by (H.1).) Not only is it natural to suppose that atoms are holeless; there is also the fact that, in principle, every hole is part of the mereological sum consisting of the hole itself together with its host. Evidently, such a sum—call it s—is not a mereological sum of the most obvious kind. Yet the peculiarity does not lie exclusively in the fact that one summand is material while the other is immaterial. The sum is peculiar insofar as one of the summands (the hole) *depends* existentially on the other. Like boundaries, holes cannot exist in isolation: they cannot be detached from their hosts. And these dependencies are not just mereological: they involve some topology too. The hole can exist only insofar as it is *attached to* its host; it must be externally connected to it (H.2).

Now, this peculiarity is important to understand how the salient parts of s are bound together. However, it does not by itself guarantee any naturalness of s. Unlike the case of self-connected material bodies, in this case there seems to be no interesting connection between the existence of a strong topological bind and the cognitive status of the ensuing mereological sum. There is no reason why the sum of a hole and its host should count as a unitary object simply because the two are attached to each other. Yet that is precisely what the negative-part theory requires. It requires that such a sum be a unitary object, for this is where the cognitive system should start when it comes to the "parsing" of holed objects. The sum is assumed to be *cognitively prior* to the analysis into object+hole. The object is a proper part of a bigger entity that includes the hole as an additional, negative part. We are prepared to accept that in some cases a holed object (as ordinarily understood) is considered a sort of incomplete object (Michelangelo's *David* with a small perforation, say). But this is not the rule. And the proof is simply that in so many cases we would not be able to tell what parts are missing from what objects.

This is what seems to be the major problem for an ontology of negative parts. We have a number of characters here. To begin with, there is an ordinary holed object, o; call it the *solid object*. Then we have the *hole, h*. The hole is not part of the solid object (which is entirely impenetrable). It is nevertheless part of the sum s consisting or the hole together with the solid object; call this sum the *holed sum*. We thus have

(8.19) $s = h+o$

and

(8.20) $o = s-h$.

Then there are two relevant mereological complements: the complement of the solid object, $\sim o$, and the complement of the holed sum, $\sim s$. The hole is part of the complement of the solid object (which for the sake of symmetry may be called the *complement sum*), but it is not part of the complement of the holed sum. For the complement sum and the holed sum mereologically overlap: they share an immaterial part—the hole. In short:

(8.21) $\sim s = (\sim o)-h$

(8.22) $\sim o = (\sim s)+h$.

Overall we thus have five main characters, three of which are obtained from the others by simple mereological composition. (See figure 8.7, which should be compared to figure 2.5 in chapter 2.) There are two more mereological mixtures, namely $(\sim s)+o$ and $(\sim o)+h$, but these play no significant role.

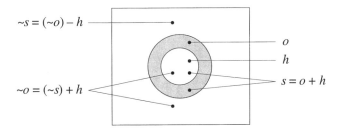

Figure 8.7
Doughnut, hole, sums, and complements.

It is enough to formulate these distinctions to see the problem emerge. A pure mereological module would founder because a negative part would at the same time be part of the holed sum and of the complement sum. The theory of negative parts can handle that. On that theory, the hole is not a negative part of the holed sum: it is a positive part thereof, and a negative part of something else. The question is: what exactly is this something else?

More characters need be added to the picture. First, in the negative-part theory we have this entity (partly solid and partly immaterial) that has h as a negative part. Call that entity o'. That is what a doughnut is, on this theory. This is not to be confused with the holed sum s because h is an ordinary part of s, not a negative part. Nor is o' to be confused with o because h is not a part of o, whether positive or negative. Rather, o' is a third object, distinct both from s and from o. Furthermore, the result of subtracting h from o' gives you yet another object, distinct from each of s, o, and o': in the case of a doughnut it gives you a disk, the disk we *would* have if the doughnut were holeless. Call this last character s'. We then have the following equations:

(8.23) $s = o + h$

(8.24) $o = s - h$

(8.25) $o' = s' \dotplus h$

(8.26) $s' = o' \eqdot h.$

The first two of these are standard mereological equations. The last two are not. The way h is added to o to yield s is not the same way h is added to s' to obtain o'; for the former operation yields a bigger object than the one we start with, whereas the latter—a form of negative sum—yields a smaller object. And the way h is subtracted from s to yield o is not the same way h is subtracted from o' to obtain s'; for the former operation yields a smaller object than the one we start with, whereas the latter—a form of negative difference—yields a bigger object. But is there any way to characterize the latter operations in terms of the former? Is there any way to characterize $\dotplus$ and $=$ in terms of $+$ and $-$? It seems not, unless negative parthood is assumed as a primitive next to parthood simpliciter. But if we do so, then we have two mereologies, not one; that is, we have two mereological primitives. And one seeming advantage of the negative-part theory over the hole theory (conceptual simplicity) is lost.

However, this is not the end of the story. Consider the complement operation. How does it behave in the presence negative parts? We are not asking about the negative counterpart of the ordinary complement operator ~, which could presumably be characterized just as ~, but using negative parthood. We are asking how ~ itself behaves when its arguments are among the additional characters envisioned by the negative-part theory. For instance, how is the following list to be completed?

(8.27) $\sim s = (\sim o) - h$

(8.28) $\sim o = (\sim s) + h$

(8.29) $\sim s' = ?$

(8.30) $\sim o' = ?$

Is h part of $\sim s'$? It would seem so, for surely h, a hole, does not overlap s', a solid disk; so it must be part of the complement. But then, what is the difference between s' and o? It can only be $s' - o$, that is, the small *solid* disk in the middle of s', which is concealed, so to speak, by the negative part of o'. However, that means that the list of entities at stake is still growing, giving rise to further questions. Call this new "invisible" part d. Is d also part of the complement of o? Of s? Of what entities? And what sort of entity is $d+h$? Finally, what sort of entity is $d\dotplus h$ (the entity obtained from d by "adding" a perfectly congruent negative part)? How does it differ from nothing at all?

These questions seem to have no answers. If the point of a negative-part theory is to avoid commitment to holes, that point is overwhelmed

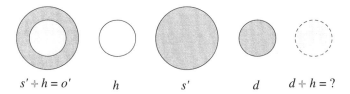

$s' + h = o'$ h s' d $d + h = ?$

Figure 8.8
Spatial entities in negative-part mereology.

by the crowd of mysterious entities generated by the theory. Better stick to our conception of holes as disjoint from their hosts: they have all the advantages of negative parts and none of their disadvantages. On the other hand, and going back to our starting point, if the purpose of a negative-part theory is to guarantee the existence of some mereological relation between the regions of two objects when one is contained in the other, then again the difficulties are overwhelming. If the region occupied by the fly is part of the region of the glass, it is also part of the region of many other mysterious characters. We may be able to express all locative relations in terms of region-location relations. But we lose track of what is located in what.

9 Spatial Essentialism

Your left and right hands are now touching each other. This could have been otherwise; but could your hands not be attached to the rest of your body? Sue is now putting the doughnut on the coffe table. She could have left it in the box; but could she have left only the hole in the box? Could her doughnut be holeless? Could it have two holes instead? Could the doughnut have a different hole than the one it has?

Some spatial facts seem tainted by necessity. But aren't spatial facts the very paradigm of contingency? To a degree, the intricate intermingling of space and modality is already visible in the part-whole structure of extended bodies. Parthood, itself a *prima facie* extrinsic relation, has an uncertain modal status. Not only is there a question concerning proper undetached parts, as we saw in chapter 6. There are also some very basic questions concerning the necessity or contingency of parthood relations. And questions about the necessity or the contingency of spatial facts and relations seem to run parallel to these.

Consider: Could an object have different parts than the ones it has? Common sense seems to have an easy, affirmative answer. At least, common sense would say that minor differences in the mereological composition of an object do not affect the identity of that object. If one molecule in this table were different, the table would not be different. If the cat had a different tail, it would be the same cat. However, this intuitive answer is deeply problematic. If one molecule does not make a difference, neither should two molecules. If n molecules make no difference, neither should $n+1$ molecules. And this is a slippery slope: unless we can pinpoint an n that does make a difference—or certain specific molecules that do make a difference, jointly or individually—we are bound to infer that the table would be the same even if all of its molecules were different. We can even imagine two tables, T_1 and T_2, consisting of exactly the same number of molecules: could T_1 consist of all the molecules of T_2 while T_2 consisted of all the molecules of T_1?

Pressed by the need to overcome such conundrums concerning the identity of mereological compounds, some philosophers have rejected the common-sense answer.[1] Most notably, Chisholm (1973, 1975, 1976) has defended the radical view that a true individual can neither gain nor lose parts, so that each single part is essential to it—a view that has come to be known as *mereological essentialism.*

> Let us picture to ourselves a very simple table, improvised from a stump and a board. Now one might have constructed a very similar table by using the same stump and a different board, or by using the same board and a different stump. But the only way of constructing precisely *that* table is to use that particular stump and that particular board. It would seem, therefore, that that particular table is *necessarily* made up of that particular stump and that particular board. (Chisholm 1973: 582–583)

This is by no means an uncontroversial way out.[2] It does, however, have some philosophical appeal and it is not in itself utterly implausible. *Strictly speaking*, if the parts change, so does the whole.

Now, regardless of how exactly one feels about these problems, if mereology generalizes to mereotopology (as we have being arguing) then it is natural to inquire about possible mereotopological extensions of essentialism. We may, for example, speak of topological essentialism to indicate the analogue of Chisholm's doctrine that results from insisting on the essentiality of an object's topology (over and above its mereology). Is the table necessarily in one piece? Would we have another table were the stump detached from the board (that very stump and that very board)? Would the table be different were some of its molecules arranged differently? Or also: Is Sue's doughnut necessarily torus-shaped? Would it be something else had its hole shrank out of existence (by rearrangements of the very same doughnut parts)? Would this chunk of Swiss cheese be a different thing had it contained one more tiny hole?

The proxies of mereological essentialism in the mereotopological domain seem to have interesting and yet unexplored consequences. As in the case of mereological essentialism, some extrinsic relations may turn out to be intrinsic relations. (The stump's being attached to the board becomes an essential property of the stump—and of the board. The hole being connected to the doughnut becomes an essential property of the doughnut.) Our purpose in this chapter is to take a closer look at these issues. Clearly this involves a decisive step into modal thinking. However,

we try to avoid excessive technicalities and rely on some intuitive understanding of the modal notions needed for our discussion.

9.1 Mereological and Topological Essentialism

Let us begin by defining the basic notions. Mereological essentialism is
not a clear-cut thesis, and various formulations have been considered. To
represent some of these formulations in a uniform way, and also to lay out
some convenient notation for comparisons with other forms of essentialism considered below, it will be advantageous to have some general way
of expressing the modal statement that a certain proposition holds in
every world in which a given entity exists. For this purpose, where ϕ is any
formula, we shall use the following general notation:

(9.1) $\Box x\, \phi =_{df} \Box(E!x \to \phi)$.

Here the symbol '$\Box$' in the definiens is to be understood as the modal
operator for necessity and 'E!' as the predicate of singular existence:

(9.2) $E!x =_{df} \exists y\, y=x$.

So (9.1) is simply a compact means for expressing the modal statement
that ϕ holds in every world in which x exists. (If desired, this notation can
also be understood in standard predicate logic terms as involving explicit
quantification over possible worlds, as suggested in section 6.2. In that
case each primitive and defined predicate would have to be thought of as
involving an additional argument and existence would itself be a binary,
world-relative predicate:

(9.3) $\Box x\, \phi =_{df} \forall w(E!xw \to \phi)$.

Here we shall conform to the more familiar notation of (9.1), which does
not require any revision in the mereotopological formalism.)

   Using this notation, we can immediately express the thesis of
mereological essentialism in terms of parthood as follows:

(PE) $Pxy \to \Box y\, Pxy$.

That is, if an entity x is part of an entity y, then x is part of y in every world
in which y exists.[3] This can acquire different meanings depending on the

properties of P. For instance, the analogue of (PE) for O yields a natural, alternative formulation of the thesis of mereological essentialism:

(OE)   Oxy → □y Oxy.

If P satisfies the Weak Supplementation Principle (P.4) of Minimal Mereology, these two formulations are equivalent. However, in the absence of (P.4) the equivalence may fail. More precisely, it is the implication from (OP) to (PE) that may fail. (In the other direction the implication is logically valid.) A counterexample is schematically illustrated in figure 9.1, which features the Hasse diagrams of two mutually accessible worlds each consisting of three objects a, b, and c (parthood relations go uphill along the lines). In both worlds, a and b overlap exactly the same things (namely, all of a, b, and c). However, contrary to (P.4), a is a proper part of b in w. Since a is not part of b in w', it follows that the model defined by these two worlds falsifies (PE). Yet (OE) holds.

It is not our purpose here to assess the viability of these different formulations of the thesis of mereological essentialism. (Chisholm himself follows (PE).) Rather, let us see how these purely mereological principles can be extended to mereotopological principles. We can immediately consider two versions of topological essentialism, corresponding to (PE) and (OE) respectively. The first can be formulated in terms of the predicate for interior parthood, 'IP' (tangential parthood would do as well):

(IPE)   IPxy → □y IPxy.

The other can be formulated in terms of the basic connection predicate 'C':

(CE)   Cxy → □y Cxy.

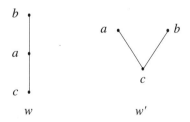

Figure 9.1
A model satisfying (OE) but not (PE).

Also in this case, it is easily verified that the two formulations are not equivalent. More interestingly, although C includes O and P includes IP there are no analogous inclusion relations between the corresponding essentialist principles. Concerning C and O, consider first two externally connected objects that are, in another world, disconnected (case (i) in figure 9.2): in this situation (OE) is vacuously satisfied, but (CE) fails. On the other hand, consider two overlapping objects that are, in another world, externally connected (case (ii)): (CE) is satisfied but (OE) is falsi-fied. Concerning IP and P, consider first an object whose internal parts become, in another world, tangential parts (case (iii)): then (PE) is satisfied but (IPE) is falsified. On the other hand, consider an object which, in some world, loses some of its boundary parts (case (iv)): then, when 'y' is inter-preted as the boundary, (IPE) is vacuously satisfied but (PE) fails.

There are also stronger (conceptually, if not logically) forms of topo-logical essentialism that one might consider. For instance, one could hold that external connection is a binding relation, in the sense that if two entities are externally connected, then they are necessarily so:

(ECE) $ECxy \rightarrow \Box y\ ECxy$.

Likewise for all the interesting mereotopological relations, such as topo-logical enclosure, tangential overlap, and the like. (See section 4.2.)

At this point the extended picture of essentialism begins to look non-trivial. There are some innocent extensions: If your left hand necessarily overlaps your whole body, then your left hand is necessarily connected to your whole body. But can your hand be disconnected from the rest of

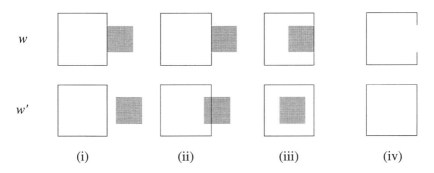

w

w'

(i)   (ii)   (iii)   (iv)

Figure 9.2
Relative independence of mereotopological essentialist principles

your body, with which it does not overlap? The relation of external connection is logically included in the relation of connection, like overlap; hence in all non-vacuous cases necessity of external connection extend to the necessity of connection. But surely one would like to say that two objects that are actually attached *could* lie apart from each other. One would like to say that the hand could be detached from the rest of the body. Thus, even if one accepted mereological essentialism, one would seem to have obvious reasons for rejecting at least some of its topological extensions.

9.2 Topological Essentialism Revisited

Well, are these reasons so obvious? What notion do we have of things that are connected? Prototypical examples are constituted by a thing and its mereological complement; or by the interior of an object and its boundary; or perhaps by a hole and its material host. In the temporal realm, typical examples are two successive events such as a process (Mary's running) and the event starting with the process's culmination (her slowing down after crossing the finish). In general, as we have seen in chapters 4 and 5, if connection is understood in the spirit of ordinary topology then external connection is a relation that can only hold between two entities one of which is closed by a boundary (in the contact area) and the other is open (in the same area). But then, in all of these cases the thesis of topological essentialism is not unreasonable. *If* it is plausible to say that the parts of an object are essential to it (as we are supposing for the sake of the argument), then it is likewise plausible to say, for example, that its complement is essential to it. This is not only obvious on a *de dicto* reading (by definition, in every world every object is connected to its complement).[4] Even on a *de re* reading the idea that the connection between an entity and its complement has the modal force of necessity (the idea, that is, that *this* object and *this* complement are connected in every world) is not in itself utterly implausible—not any more implausible than mereological essentialism.

As a matter of fact, one can even argue that this thesis is *implied* by mereological essentialism, as long as certain assumptions are made concerning the logic of parthood. For consider a world w in which y is the complement of x. If in another world w' it turns out that y is no longer connected with x, then the Weak Supplementation Axiom (P.4) implies that there must be a part of the universe, z, that belongs neither to x nor to y. By (PE), z must be a part of the universe in w too. Therefore, it must overlap either x, or y, or both. But if it does, it is a contingent fact, as shown by w', and this is ruled out by (OE). This argument shows that in the presence of (P.4), the thesis that everything is necessarily connected to its complement—in its *de re* reading—follows from (OE) together with (P.4) and the rather innocent assumption of the existence of the universe.

The cases in which the principles of topological essentialism do sound implausible arise when the things whose contact is said to be necessary are ordinary objects, such as a book and the shelf it stands on, or a painting and the wall it hangs on. There is indeed a lot of *prima facie* implausiblility in the statement that these things are necessarily touching—that the book could not be on a different shelf, or that the painting could not be hanging on a different wall. There is a lot of implausibility in the thesis that the book or the painting would literally cease to exist or no longer be what they are upon removing them from the shelf or the wall with which they are presently in contact. But on closer inspection, these are not counterexamples to topological essentialism, in any of its forms. They are not counterexamples because the relevant relation of contact here is not one of topological connection. The book and the shelf, or the painting and the wall, are not externally connected. They are, as we may say, quasi-connected (QC). They are very close; but as we have seen in section 5.8, closeness is no mark for connection.

We could, of course, express a correspondingly strong essentialist principle to the effect that any two things that are quasi-connected in this loose sense are so in all worlds in which they exist.

(QCE) $QCxy \rightarrow \Box y\, QCxy$.

However, this is no innocent topological extension of mereological essentialism. It is a much stronger and independent, substantive thesis—a form of *metric* essentialism whose rebuttal would not affect the question of the plausibility of topological essentialism, and which therefore does not have much bearing on the question of the relationship between mereology and topology. Our initial example of your two hands touching also falls into this case. In any event, even in this regard one could find some plausibility in the essentialist thesis, at least relative to certain kinds of entity. Would that mosaic be the same if the tesserae were arranged differently? Would that Tinkertoy house be the same if the Tinkertoys were arranged in totally different way? Would this beautiful bunch of flowers be the same if the flowers were scattered all over the floor?

There is, finally, the case of contact between two parts of a given object, such as your hand and the rest of your body, or two halves of a solid sphere. They are connected, and we know this is a case of strong connection: their sum is strongly self-connected. Are they necessarily so? Here it seems that the intuitive plausibility of an affirmative answer would indeed mark a much stronger commitment than mereological essentialism. It seems quite plausible to maintain that the two halves are essential to the sphere without entertaining the additional view that they cannot be separated—that the sphere cannot be cut in half. We agree with this. But, once again, things look different on closer inspection. Surely the two parts can be detached (no essentialist thesis would deny that the sphere is *divisible*). The question is whether they would survive the separation—whether the connected halves are the same things as the disconnected halves. Equivalently, the question is whether the sphere survives the cut—whether the connected sphere and the split sphere are one and the same thing. It would seem natural to answer these questions in the affirmative. After all, *nothing* happened to the two halves, except for a change in their extrinsic, relational properties. And how could a mere extrinsic change result in a substantial change?[5]

However, this is hasty if our concern is with topological facts, for topological relations need not be of the extrinsic variety. The splitting transforms the two halves into full-fledged, maximally connected things. Two entities that were only partly bounded by a surface now are completely bounded and perfectly separated from their complement. Or, if you prefer, two merely potential entities have become actual. On either view, the change is remarkably substantial: the parts remain the same, but the boundaries change. And to say that nothing happened, to deny that the two halves undergo a dramatic change, is to beg the question. If a mereologist can be struck by microscopic mereological changes, a mereotopologist has all the reasons to be struck by such macroscopic topological changes.

These considerations suggest that topological essentialism may be stronger but in no way stranger than its pure mereological counterpart.

The last case is the only one where one would need additional grounds to argue from mereological essentialism to the thesis that the two halves of a sphere cannot be separated. But even so, this brief discussion shows that such additional grounds are a rather natural extension of whatever reasons one could offer in favor of mereological essentialism.

9.3 Spheres and Doughnuts

There are other senses in which topological essentialism may not seem unreasonable, for there are other forms of topological essentialism besides those expressed by modal statements such as (CE), (IPE), (ECE), and the like. These statements express a clear sense in which the essentialist position can be extended from mereology to mereotopology by insisting on the essentiality of such characteristic relations as interior parthood, tangential parthood, etc. However, one can also think of the essentiality of topological properties of a different sort, such as the properties as being topologically sphere-like, doughnut-like, and so on—in short, those properties that determine to the topological *genus* of an object. Correspondingly, one can formulate principles of topological essentialism that force the genus of an object to be the same in every possible world. This is the sort of essentialism that arises in relation to our initial questions about Sue's doughnut.

There are, again, *de dicto* and *de re* formulations of such principles. Writing 'Dx' for 'x is a doughnut' and '$PHyx$' for 'y is a perforating hole in x' (see 8.15), we could express the *de dicto* formulation as follows:

(DD) $\Box \forall x(Dx \rightarrow \exists y\, PHyx)$.

This seems uncontroversial (if not analytically valid).[6] Surely you cannot have a holeless doughnut, for having a hole is definitory of being a doughnut: in every world in which you have a doughnut, you have a hole in it (just as in every world every husband has a corresponding wife.) There is nothing peculiar about topology here, and there are perfectly comparable forms of *de dicto* mereological necessity: surely you cannot have a sphere with the right half removed. A truncated sphere is simply not a sphere.

The interesting questions arise with the *de re* formulation. Given a doughnut, is it true of it—of *that* particular object—that it could not exist without its hole—*that* very particular hole? To be sure we can specify this claim in at least two different ways:

(DE.1) $Dx \rightarrow \Box x \, \exists y \, \mathrm{PH}yx$

(DE.2) $Dx \rightarrow \exists y \, \Box x \, \mathrm{PH}yx.$

The first of these is a rather weak claim: it says nothing about whether a given doughnut could have a different hole from the one it actually has. The second is a much stronger claim and rules out the possibility that a doughnut could be perforated by a different hole: in every world in which you can find Sue's doughnut, you will also find the very same hole that it has in this world. Both claims of course have various generalizations, concerning for instance the number of holes in an object and eventually also its topology (a doughnut with a knotted hole). Here we may content ourselves with the simple cases.

Is either of these theses acceptable? On one intuition they are both too strong. If you open up a doughnut (e.g., by cutting it on one side) the hole goes, the doughnut stays. That is, the *object* is still there, and you can still eat it all, though its shape is now different. This would contrast both (DE.1) and (DE.2). On the other hand, there is no a priori reason why this commonsensical intuition should be regarded as a *reductio ad absurdum* of either (DE.1) or (DE.2). Concerning (DE.1), there is actually a close connection between this and the topological essentialist theses expressed by (CE) and (ECE). If you open up the doughnut, some parts that were connected become disconnected. So if (CE) or (ECE) hold, (DE.1) must hold too. (However, the converse need not be true, hence (DE.1) is effectively less committing than the other theses.) Moreover, there is also a connection between (DE.1) and the mereological essentialist theses expressed by (PE) and (OE). The hole in a doughnut is part of the doughnut's complement, and a tangential part at that. Thus, to the extent that the relation of external connection between every object and its complement is a *de re* necessity that follows from mereological essentialism (as seen in section 9.2), to that extent the relation of external connection between a doughnut and its hole is itself a matter of *de re* necessity supported by mereological essentialism. So if (PE) and (OE) hold, (DE.1) must hold too.

These considerations do not extend to (DE.2). Yet also in that case reference to common sense and intuitions is hardly a way to settle the

Figure 9.3
Opening up a doughnut kills the hole in it. Does the doughnut survive?

issue. If the thesis that each part of the doughnut is essential to it can survive the cry of common sense, so may the thesis that the hole—that very hole—is also essential. Actually, if the theory of negative parts were right, then a conspicuous part of the story here would reduce to plain mereological essentialism. This hole would be necessarily in that dough-nut insofar as the hole is a part of the doughnut—a negative part. Rather than an advantage, however, we take this to be a further element against that theory: there is an important distinction here—a distinction reflected in the possibility of endorsing topological essentialism without endorsing mereological essentialism, or vice versa. And this distinction is obscured by a negative-mereological treatment of holes.

Incidentally, each of (DD), (DE.1), and (DE.2) was phrased in terms of the binary predicate 'PH' introduced in chapter 8, which reflects a reifying attitude towards holes. However, we can also imagine an expres-sion of the form '$\exists y\ \mathrm{PH}yx$' to be a metalinguistic abbreviation for a complex expression attributing to x the property of being perforated, in the spirit of the adjectivalist account discussed in section 2.3. On such an account, the possession of a hole would simply be a *façon de parler*, and 'there are holes in' would be an innocuous shape predicate like 'is a doughnut'. This has no consequences for the *de dicto* principle (DD), which in fact becomes synonymous with the statement that necessarily a doughnut is a doughnut. However, the status of the *de re* principles changes on this account. For one thing, the strong formulation (DE.2) becomes meaningless (syntactically ill-formed). As for (DE.1), the thesis remains intelligible, but it becomes an instance of a more general thesis to the effect that an object must have the shape it has. If sameness of shape is understood topologically (i.e., modulo topological equivalence), then the result is equivalent to (DE.2), and bears the same relations to the other essentialist principles of mereology and topology that were noted above. If, by contrast, sameness of shape is taken literally, then the result is a much stronger thesis, which may be labeled *morphological* essentialism. Can an object have a different shape than the one it has? Does it cease to exist when its shape changes? Whether this is metaphysically acceptable, or *when* it is acceptable if at all, is a new question—another step into the intricate intermingling of space and modality.

9.4 Locative Essentialism

We can go even further in that direction. Consider strengthening the link between an object and the region of space at which it is located by asserting the necessity of location. This yields a form of *locative* essentialism which is stronger than any form of spatial essentialism considered so far, and can be formulated as follows:

(LE) $Lxy \rightarrow \Box x\, Lxy$.

That is, an object that happens to be located at a certain region is necessarily located there, whenever it exists. Other versions, exploiting the various locative relations of chapter 7, could of course be added to the list. The generating scheme is

($\mathcal{R}$LE) $\mathcal{R}Lxy \rightarrow \Box x\, \mathcal{R}Lxy$,

where '$\mathcal{R}$' is any locative modifier (so that $\mathcal{R}L$ is any relation among GL, WL, PL, IWL, IPL, TWL, TPL, and possibly other relations as well).

These principles imply that no object can survive with impunity a rearrangement of its parts. This is stronger than mereological essentialism (PE) and topological essentialism (CE), since a rearrangement of parts need not carry a loss of parts (violating (PE)) or a change in the topology of the object (violating (CE)). For instance, in figure 9.4 one can observe a delocation of parts with no mereological or topological change. This would falsify (LE) while being compatible with both (PE) and (CE).

Now, like their mereotopological analogues, all these locative theses express *de re* modalities. There are, to be sure, weaker *de dicto* versions of locative essentialism, which can be formulated in terms of the region operator r. For instance, the following gives a *de dicto* counterpart of (LE),

(L) $\Box \forall x \forall y (rx=y \rightarrow Lxy)$,

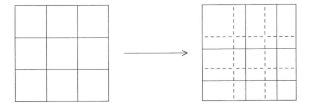

Figure 9.4
(PE)–(OE) and (CE)–(IPE) are preserved; (LE) fails.

and this is obvious. Surely *x* is located at *y* whenever *y* qualifies as the region at which *x* is located. This is the same sort of triviality that we encountered in the *de dicto* versions of the other principles. It is the corresponding *de re* forms that are questionable. Presumably the painting that is now on that wall could have also been on this wall. To deny this seems to be an intolerably severe position—certainly much more severe than the position expressed by the above principles of topological essentialism, let alone mereological essentialism. We are not aware of any philosophers who ever held such views. But consider, for an analogy, van Inwagen's words about the temporal case:

> It is bad enough to suppose that the replacement of a rusty bolt leaves me with what is, "in the strict philosophical sense," a new car. It is infinitely worse, and never had the phrase 'infinitely worse' been used more appropriately, to suppose that when I sit in my car and turn the wheel, what I am occupying is, "in the strict philosophical sense," a compact series of infinitesimally differing cars. (1990: 77f.)

This statement can be rephrased to match our concerns. It is indeed bad to suppose that, had the wheel of the car been three degrees more to the left than it actually is, the car would have been another entity.[7]

Similar considerations apply, it seems, to each instance of (𝓡LE) as also to various other forms of locative essentialism that could be formulated, thanks to the many distinctions afforded by the basic mereotopological vocabulary. For instance, one could consider weakenings of (𝓡LE) to the effect that if an object is located (exactly, partly, wholly, etc.) where another object is, then the two are always going to bear that spatial relationship to each other (though they might slightly change their exact location). In the terminology of chapter 7, this can be expressed in terms of suitable region-location predicates. The generating scheme is

(R𝓡LE)   R𝓡Lxy → □x R𝓡Lxy.

In the terminology of chapter 8, the relevant predicates express containment and the generating scheme is

(𝓡INE)   𝓡INxy → □x 𝓡INxy.

Statements of this form are generally weaker than the principles defined by (𝓡LE). But van Inwagen's misgivings would still apply. It is hard to imagine that the car you are sitting in would be another thing had the region it occupies not been, say, an interior part of the region occupied by the garage. It is hard to imagine that the fly that is in the glass would have been another thing had it not been in the glass.

   Is there any use for these forms of essentialism, then? Note the difference between (R𝓡LE) and the forms of topological or metric essentialism exemplified by (CE) and (QCE). Those are principles concerning the relative positions of certain objects: (CE) demands that two halves of a sphere be necessarily in contact with each other, but leaves room for the possibility that they jointly move around in the environment; (QCE) demands that the book be necessarily on the shelf, or that the tesserae of the mosaic be necessarily arranged as they are, but leaves room for the ideal possibility that the book be moved along with the shelf, or that the mosaic be somewhere else (insofar as this does not collide with other essentialist facts). By contrast, (R𝓡LE) concerns the relative position of spatial regions. Now, on a *de dicto* reading, this is much more exorbitant than any of the above: (R𝓡LE) becomes a thesis to the effect that things cannot be elsewhere—whether individually or together with other entities in the environment, a thesis beside which even the strongest forms of mereological and topological essentialism pale. However, on a *de re* reading (R𝓡LE) is commensurable to the corresponding forms of topological or metric essentialism. In fact, insofar as spatial regions are entities of a kind, (R𝓡LE) says neither more nor less that those entities are necessarily 𝓡-related, which is what (CE) and the like say. (A relationalist about space could hold analogous views, though the actual content of the principles would be different. Specifically, one can obtain relationist analogues of the principles of locative essentialism by replacing each occur-

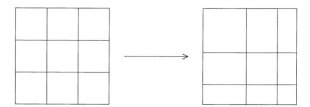

Figure 9.5
In a spatial domain, (LE) supervenes on (PE)–(OE) and (CE)–(IPE).

rence of a binary predication 'Lxy' with the idiom 'L*x' that gives its relationist translation.)

On a *de re* reading, then, (R𝓡LE) is not an independent thesis. Just as mereology forces certain topological facts, so topology forces certain locative facts—those facts that concern the location of spatial regions. This yields an asymmetry between (R𝓡LE) and (LE), since (LE) was seen to be independent of both mereological and topological essentialism. It does not follow, however, that (LE) is unintelligible. Consider a purely spatial model—a model in which every world is inhabited exclusively by regions of space (including Space, the fusion of all these regions). Assume both mereological and topological essentialism, in any of the forms discussed above. Then the corresponding instances of (R𝓡LE) are true. But then (LE) must also be true. For this principle states that regions cannot be located elsewhere. So how can it be violated, if the only interpretaton of this 'elsewhere' is given in term of location itself?

There is a holistic character in the decomposition of space into regions: In a purely spatial domain, the difference between the two chessboards in figure 9.5 is merely decorative, for regions cannot be anywhere else than where *they themselves* are. In a purely spatial domain, the distinction between *de re* and *de dicto* readings of (LE) collapses, and the entire essentialist story supervenes on matters of mereology and topology.

10 Events in Space

Our account of the structures of spatial representation so far reveals a tension. On the one hand, we have stressed the importance of a pluralistic ontology, at least as a methodological stance, suggesting that spatial entities of different types may bear different relations to space. The analysis of location in chapter 7 was meant to substantiate that hypothesis. On the other hand, we have endorsed a way of representing spatial features (in terms of parthood and connection relations) that draws no significant distinction between entities of different types. We have been speaking of parts and boundaries in an abstract fashion, following Husserl's plea for generality. And although this is to some extent a necessary and quite convenient move for the purpose of a formal account, the fact remains that these notions may not be equally appropriate or share the same characteristics in all cases.

The opposition between objects and events is most indicative of this tension. We have seen that ordinary objects, such as tables and stones, are tightly and clearly connected to space. They have an exact spatial location and, indeed, they are the sole *occupants* of the spatial regions at which they are located. Their mereotopology is spatially clear too: the proper parts of a table occupy corresponding proper parts of the region occupied by the table, and they are connected to one another just in case the corresponding regions are connected. So tight and clear is the relationship material objects bear to space that they are the paradigm of spatial entities. Ordinary material objects have also a tight, but less clear, connection to time. For one thing, they *change* through time: they have different properties at different times, and may bear certain relations to other objects at some times but not at others. (For instance, they can move, unless we accept some form of locative essentialism.) Moreover, material objects have beginnings and endings: their life-spans (their *histories*) have an initial point and an end point, though these points might be hard to determine.

The connection to time is, by contrast, very determinate in the case of events—those things that *occur* or *take place*, such as picnics, explosions, whistles, runs. Events are embedded in time. They have temporal boundaries. So tight and clear is their relationship to time that the simplest criterion for individuating an instant or period of time consists in providing a (real or possible) event that takes place at that time. (Thursday is when John came for dinner. Mary called just when the movie was about to start.) The conceptual connection between such temporal entities and space is, however, less clear. Events, we said, do not occupy space, unlike material objects. But how, exactly, are they related to space? Where do they take place? How are their parts spatially related to one another? One might be tempted to answer these questions by refererring to the spatial locations of the material objects involved in the event—the "participants" of the event. For instance, Davidson (1969) and Lombard (1986) have suggested that an event's exact location is the location of the minimal participant of the event. At each instant of time, the exact location of John's raising of his arm is the spatial region occupied by John's arm. Since the same participant, or the same part of a participant, can be involved in more than one event at the same time, events cannot on this account occupy the space where they occur. Yet we would have a clear criterion for their spatial location. On the other hand, Quinton (1979) and Hacker (1982) have objected that the question 'Where did . . . happen?' is rarely a question about a minimal area in this sense, and may actually be meaningless for various classes of events (e.g., for most social events or purely relational changes). The notion of change, too, is problematic in the case of events. Surely the play was boring at the beginning and got exciting at the end. But this means that the first part of the play was boring, and the last part exciting. Events have *temporal parts* (phases) and these can have different properties. Is this a way of saying that events can change? And can events move?

These questions indicate indeed the existence of deep disanalogies between events and objects with regard to their spatial and temporal dimensions. However, the spatio-temporal analogies that these questions delineate are important too. For instance, the basic concepts—such as the concepts of part and boundary—appear not to be domain-specific. There can be spatial as well as temporal boundaries and, more importantly, structural analogies between these two sorts of boundary suggest that 'boundary' is not used in different senses in the two cases. Likewise for parts. We know from the beginning that parthood is not specifically a spatial concept; but the existence of structural analogies between its behavior in the spatial and temporal domains is a fact well worth investigating. We are also helped here by some linguistic evidence: a range of predicates can be applied rather indifferently to both objects and events. Think of verbs such as 'begin', 'end', 'spread' (as in 'Italy/the performance begins here', or 'the tree/epidemic is spreading all over') or prepositions such as 'in', 'at', 'through', and 'over' (as in 'You will be *at* the station *at* 3 P.M.' or 'Jean moved *over* three countries *over* the last five years, and *in* these last months she decided to live *in* Venice'). There is no immediate explication for this relative topic-neutrality. But it is noteworthy because it suggests the existence of a deep connection between spatial and temporal characterizations of objects and events.

This connection is the focus of this chapter. We think that the spatio-temporal analogies mentioned above may play an important heuristic role in the investigation of spatial concepts vis-à-vis their temporal cognates. And we argue that their consideration allows us to import some rather clear results about the temporal dimenson of objects into the much more unfocussed problem of the spatial dimension of events.

10.1 Spatio-Temporal Analogies

The philosophical study of spatio-temporal analogies and disanalogies has two main sources. On the one hand, there is the metaphysical quarrel about the nature of objects and events. According to one line of thought, tokened by Russell (1927) and Whitehead (1929) and championed by Quine (1950, 1960, 1985), the analogies between objects and events are so strong that the two categories may in fact be one. In Quine's words:

Physical objects, conceived . . . four dimensionally in space-time, are not to be distinguished from events or, in the concrete sense of the term, processes. Each comprises simply the content, however heterogeneous, of some portion of space-time, however disconnected or gerrymandered. (1960: 131)

What then distinguishes material objects from events is, on this view, not a matter of kind but one of degree: an object is more stable than an event; it is "fairly firm and coherent internally, but coheres only slightly and irregularly with its spatio-temporal surroundings." (Quine 1970: 30)[1] As a reaction to this view, "descriptive metaphysics"[2] philosophers such as David Wiggins have urged that "we take note of all the hints of analogy and disanalogy we get from the unreformed language of things and events." (1980: 26, n12) And for Wiggins the disanalogies are dominant:

A continuant [an object such as a table] has spatial parts, and to find the whole continuant you have only to explore its boundary at a time. An event has temporal parts, and to find the whole event you must trace it through its historical beginning to its historical end. An event does not have spatial parts in any way that is to be compared with (or understood by reference to) its relation to its *temporal* parts. (ibid.)[3]

On the other hand, there exists a tradition of studies on the logical possibility of time travel that purports to show (or to criticize the idea) that on a suitable interpretation time travel is equally plausible as space travel.[4] In this context, Richard Taylor has held that temporal and spatial relations are very much alike—more precisely,

that (1) terms ordinarily used in a peculiarly temporal sense have spatial counterparts and vice versa, and that accordingly (2) many propositions involving temporal concepts which seem obviously and necessarily true, are just as necessarily but not so obviously true when reformulated in terms of spatial relations; or, if false in terms of spatial concepts, then false in terms of temporal ones too. (1955: 599)

It is part 1 of this program that concerns us here. We do not wish to enter the quarrel about the nature of events, just as we aim at some neutrality concerning the nature of material objects. But we are interested in the idea of a structural similarity between objects and events insofar as this may help sharpen our understanding of the basic structures involved in spatial representation. Let us say that a characterization of the category of objects or the category of events is a *structural* characterization if it hints exclusively at spatial or temporal facts. Now suppose that, given a structural characterization of objects, you can produce a structural characterization of events simply by replacing temporal facts (if any) by spatial facts, and spatial facts (if any) by temporal facts. And suppose, conversely, that given a structural characterization of events you can by a similar replacement produce a structural characterization of objects. (In a weak form, replacements produce structural characterizations of an as yet unidentified type of entities, which might well turn out to be objects or events.) This possibility is the content of what we shall call, following Mayo (1961), a complementarity hypothesis, and it is with this hypothesis that we shall work. Are there any adequate structural characterizations of objects and events that satisfy the complementarity hypothesis?

10.2 Extension and Duration

Mayo (1961: 343) draws a detailed list of the spatio-temporal principles that articulate the complementarity hypothesis. Here is a first pair:

(O1)   An object (a) has a limited *extension* and an unlimited *duration*; (b) it cannot take up the whole of *space*, but it could take up the whole of *time*; and (c) there must be room in *space* for many *objects*, which may or not overlap *temporally*.

(E1)   An event (a) has a limited *duration* and an unlimited *extension*; (b) it cannot take up the whole of *time*, but it could take up the whole of *space*; and (c) there must be room in *time* for many *events*, which may or not overlap *spatially*.

These are structural characterizations. They appear to satisfy the complementary hypothesis, as indicated by the italicized words. And both seem to us acceptable, provided of course that we take the notion of "taking up" space and time with sufficient flexibility, without implying occupation in the sense of section 7.7.[5] To be sure, there is some ambiguity in the notion of an object or event having 'limited' extension and duration, respectively. It will presumably depend on the topology of space and time whether there can be such things as an unlimited object (an indefinitely extended blob) that does not extend *everywhere*, or an unlimited event (a whistle) that does not last *forever*. But these are details that can be fixed without affecting the spatio-temporal analogy between the two characterizations. We can certainly agree that, *in the norm*, paradigm objects and events are bounded entities.

   To see, however, whether these characterizations are an adequate instantiation of the complementarity hypothesis, we have to scrutinize more closely in what sense the propositions in (E1) are derived from those in (O1) "by exchanging spatial for temporal terms and vice versa," as Mayo puts it. Take proposition (a). We are asked to regard 'duration' and 'extension' as complementary temporal and spatial terms. This can be easily seen if we read 'extension' as 'spatial extension' and 'duration' as 'temporal extension'. That 'spatial extension' is complementary to 'temporal extension' is now straightforward, since 'spatial' complements 'temporal'. A similar redefinition should apply, say, to 'size' and 'length', where the latter is meant to be just temporal size. If we succeed in establishing these propositions, then the complementarity of (O1) and (E1) becomes interesting.

Here, however, our charitable reading of (a) causes problems. If the (a)s are really about paradigm objects and events, as we must suppose, then the rewriting produces results that do not seem to respect the conceptual flavor associated with the source. For talk of the unlimited duration of an object and of the unlimited extension of an event is no more felicitous than talk of the limited extension of an object and of the limited duration of an event. Unless we stipulate otherwise, *paradigm* objects such as glasses and chairs have not too long a life, and events such as picnics, big as they might be, do not *normally* trespass certain very obvious limits. This seems enough to undermine the analogy between (O1) and (O2).

10.3 Motion and Location

Consider a second spatio-temporal analogy offered by Mayo in support of the complementarity hypothesis:

(O2)   An object can, at different *times*, take up the same *space* (rest, endurance) or different *spaces*, normally of the same *size* (locomotion, endurance).

(E2)   An event can, at different *places*, take up the same *time* (occurrence, extension) or different *times*, normally of the same *length* ("propagation," extension).

Again, (O2) and (E2) seem to satisfy the complementary hypothesis. As a matter of fact, this seems to be more than an analogy: the results of the exchange are not only symmetric, but virtually indistinguishable from their sources. It just does not matter much if we talk of being at different spaces at different times or of being at different times at different spaces. Yet, plausible as the analogy may seem, two problems arise. First, we are told that objects move and that events propagate, but what is the difference between *locomotion* and *propagation* (i.e., what is the proper meaning to be assigned to 'propagation', insofar as it is supposed to complement 'locomotion'), apart from the fact that the former is predicated of objects, and the latter of events? Second, does this difference depend upon spatial or temporal facts only? If 'event movement' is to be defined in terms of the structural complement of object movement, then no apparent asymmetry emerges: event movement and object movement are one and the same thing.

Now, can events move? A negative answer comes from Dretske (1967). Dretske acknowledges that some events may occur in spatial containers, which can move. However, this does not allow us to say that the events themselves move, for events and objects are at their spatial location in different ways. A chair can be said to be in a building (at a particular time) even though most of its life is spent elsewhere. A picnic cannot be said to occur in a building if it just starts there but then winds up in the garden; we should say, rather, that some initial *part* of the picnic was in the building and a later *part* was in the garden. (This is in the spirit of Wiggins's remarks to the effect that objects, but not events, are wholly present at any time at which they exist.) At most we can say that the picnic occurred in the place that is the mereological fusion of the space of the building and the space of the garden, or of the relevant portions thereof. Either way, there is an incompatibility between our ascription of spatial location to events and the concept of movement, which Dretske characterizes as follows:

(D)   For anything to move it has to be wholly present at different places at different times.

So if events can at most be partly present (i.e., located) at different places at different times, it follows that they cannot move.

Hacker (1982) adds that the difference between objects and events regarding movement is apparent from their respective relation to space, namely, that objects do and events do not occupy (exclusively) their location. However, we do not think that the notion of occupation should play any relevant role here. One can accept the fact that events do not occupy space and also the fact that events do not move without accepting that there is any logical relationship between these facts. We may assume that if something occupies a region r, it is wholly located at r (more precisely, it is wholly located at r or at some larger region r').[6] But we know that the converse is not true. Hence from the fact that something does not occupy r one cannot infer that it is not wholly located at r (or at some extension r'). The immobility of events seems to be linked only to the fact that events cannot be wholly located at different places at different times.

Let us, then, take a closer look at (D). If this intuitive criterion for motion is taken literally, then all motion involves a change of location over time. However, there are clear cases of motion of material objects for which this is false. Consider a perfectly circular disk spinning around its axis: it moves; but it does not undergo any translational motion. The starting place and the landing place of the disk, and every intermediate place too, are one and the same. More generally, all sorts of perfectly circular motion provide counterexamples to (D).[7] The point is not only epistemological, i.e., it does not merely concern the impossibility of telling a spinning disk from a still disk (we can always paint stripes on the disk in order to see whether it rotates). The point is that circular motion is perfectly normal (there is nothing metaphysically bizarre about it) and yet (D) makes circular motion impossible.

There is an obvious improvement of (D). We can require that at least *some parts* of the moving object change their location:

(D′)   For anything x to move, at least some parts of x have to be
         wholly present at different places at different times.

Obviously, objects that move translationally satisfy this improved version just as they satisfy (D): *all their parts* change location. But now the spinning disk would be moving too, for some of its parts are now here that will be at another time there. (An interesting consequence of (D′) is that mereological atoms could translate but could not rotate.)[8] Observe, here, that the rotation of circular objects provides yet another illustration of the central role of mereology in spatial thinking. One can tell (and describe) the difference between the disk that is rotating and the one that is still only if one can distinguish between (and speak of) parts and wholes.

Unfortunately, the main consequence of our emending (D) as (D′) is that the result does not yield any significant distinction between objects and events—which is what (D) was supposed to do in the first place. Maybe there is no picnic, and no event in general, all parts of which are first in one place, and then in another. But the picnic that starts in the building and lands in the garden has some parts in the building and some other parts in the garden. Thus, the picnic moves—at least in the same sense in which the disk rotates.

One asymmetry seems to remain, though. Disks rotate, but does the same applies to picnics? In the middle of a football game there is a rotation—players exchange fields. But is this a rotation of the game? Dancers in a ballroom traced more or less circular trajectories, but did the second waltz of the evening itself rotate? The rotation here is more plausibly predicable of the plurality of actors in these events—the teams, or the group of dancers—but not of the events themselves.

One way of explaining the conceptual asymmetry here would be to reject the notion of a spatial part of an event altogether. For on all other aspects the notions involved in describing events and, respectively, objects appear to be conceptually symmetric: (D') applies directly to both objects and events. One might thus think that symmetry founders *on the very notion of part*. For the parts of the picnic that are now here and then there are temporal parts, or phases of it; whereas the parts of the disk that are now here, and then there, are spatial parts. One might therefore object that it is spatial parts that are required to move according to (D'), and that it is the notion of a spatial part of an event that is unclear. Peter Simons, for one, would not agree with this—events have spatial as well as temporal parts:

> In a football match, the first half of the play is a temporal part, the events taking place in one half of the field make up a spatial part, and the part played by one of the players is neither purely spatial nor purely temporal. (1987:131)

Still, the difficulty remains to see why spatial parts play an important role in accounting for the motion of objects and play no clear role in the disputable case of event motion.

10.4 Imperfect Complementarity?

If we agree with Simons that events have spatial parts (such as the left half of a football game), why shouldn't we also accept the complementary thesis that objects have temporal parts?

One might think that there are independent objections to the doctrine of temporal parts of objects. It might be thought that the problem with the temporal parts of an object is that it is difficult to see what sort of properties they could have. What is the shape of the temporal part of John in the period in which he turns around a table? Is it a toroidal shape, with the table occupying the hole in its middle? How is such a shape to be related or contrasted with the human-like shape we believe John has? Or, suppose John is happy for the most of his life, and unhappy during a very short phase of it—can we use this fact for classifying John's temporal parts into the short, unhappy one and the long, happy one? But obviously such odd problems arise with spatial parts of events as well. If football were played in triangular fields, such that one team at a time would play in a trapezoidal "half," would we talk of the trapezoidal part of the game? If the grass were greener on the left half of the field, would we talk of the green part of the game? Judith Thomson (1971) says that objects with temporal parts yield a "crazy metaphysics," and she may well be right. Does this imply that a metaphysics including events with spatial parts is crazy?

Again, the matter cannot be one of mere stipulation; for if it were, then nothing would prevent one from taking the Wiggins-Simons characterization of objects:

(O3)   Characteristic of an object is that at any time when it exists, it is present as a whole, and not just in part.

and produce a complementary characterization of events:

(E3)   Characteristic of an event is that at any place where it exists, it is present as a whole, and not just in part.

The complementary characterization (E3) would rule out spatial parts of events exactly as (O3) ruled out temporal parts of objects. Now the reason adduced by Simons (1987: 130) for accepting (O3) is this:

(O4)   When an object has first one property and then another, contrary property, it is the whole object—not different parts of it—that has these properties. By contrast, with events we can always refer such temporary properties down to temporal parts.

But, again, look at how sensible the complementary version of this characterization sounds:

(E4)   When an event has here one property and there another, contrary property, it is the whole event—not different (spatial) parts of it—that has these properties. By contrast, with objects we can always refer such local properties down to spatial parts.

Our wording of (E4) shows what is wrong with (O4): it equivocates on 'part'. If 'part' is understood as spatial part, the thesis is obviously true; but if it is understood as temporal part—as required by the complemen-

tarity transformation—then (O4) cannot be used to argue against existence and relevance of such parts.

To take stock, there seems to be no important difference between
objects and events as to the possession of parts of any kind. There might
be differences between the ways in which events relate to their spatial
and temporal parts respectively; but, for that matter, one can find quite
analogous differences between the ways in which objects relate to their
temporal and spatial parts too. There seems to be no deep asymmetry
here between objects and events, but at most an asymmetry between the
relations events and objects bear to their own parts. All we can accept, for
the time being, is the weaker thesis expressed by Wiggins in the passage
quoted above:

(E5)   An event does not have spatial parts in any way that is to be
       compared with (or understood by reference to) its relation to its
       temporal parts.

This leaves room for the possibility that events have spatial parts. And we
can reasonably extend (E5) to a similar claim about objects:

(O5)   An object does not have temporal parts in any way that is to be
       compared with (or understood by reference to) its relation to its
       spatial parts.

We are not saying that the notion of a temporal part of an object is to be
recommended. Simply, from a structural point of view it is neither more
nor less plausible than the notion of a spatial part of an event. Complementarity is here to stay.

10.5   Types of Movement

Let us summarize our results and complete the picture. We have the
following theses:

(1) In order for an object o to move, its spatial parts must be located at
different places at different times (this is principle (D')).

(2) In order for o to rotate (without translation), its spatial parts must be
located at different places at different times, but the whole of o must be
always located at the same place, and the parts must bear the same spatial

relations to each other (barring this conditions would yield a description of a sort of internal viscous movement).

(3) In order for o to translate, it must be wholly located at different places at different times.

(4) In order for o to expand, some of its spatial parts must be located at different places at different times, but some other parts must remain at the whole of the place that o occupied at the beginning.

We have also accepted the complementary of (1):

(1') In order for an event e to move, its temporal parts must be located at different places at different times.

But intuitions are unclear as to whether we can make finer distinctions for event movement—whether we can speak of event expansion, rotation, or translation.

Now, the lack of a complete correspondence here does not impinge on the facts of complementarity. The only conclusion we are allowed to draw is that event movement is not complementary to object movement. Second, observe that according to (2) and (3) an object cannot rotate and translate at the same time. This is only superficially counterintuive. Even though we can decompose an object's movement into a rotation and a translation (if, for instance, the object describes a cycloid trajectory), it does not follow from this that the object in fact both rotates and translates. As a matter of fact, an object's trajectory can be described as the resultant of indefinitely many components on indefinitely many trajectories, and it does not follow from this that the object moves in indefinitely many different ways. On the other hand, it is fairly obvious how to modify the previous propositions so as to cover the case of compound kinds of movement (translation *cum* rotation, translation *cum* expansion, rotation *cum* expansion, etc.).

10.6 More Analogies

Let us, finally, consider a third set of analogies offered by Mayo in support of the complementarity thesis:

(O6) An object cannot be at different places at the same time, unless its spatial size is greater than the interval between the places.

(E6) An event cannot be at different times at the same place, unless
      its temporal size is greater than the interval between the times.

These two principles rely again upon the interaction between spatio-temporal and mereological concepts, and again, they require some qualification. First of all, (O6) excludes that, say, one single normally sized
one-piece chair be located simultaneously in Paris and New York. A chair
can be located in both places in two circumstances only: either it is at
different times in these places (it moves thus from the one to the other)
or it is simultaneously at both places, but only partially so. In the last case,
there are two quite different possibilities: (i) the chair is so big that it
reaches Paris and New York and has, say, one leg here and another there;
(ii) the chair is a normally sized chair, but it has one leg in Paris and
another in New York because it is broken into two pieces—it is a scattered chair. (This is a possibility not contemplated by Mayo; we ought to
assume that Mayo understood wholes as self-connected.)

The complementary case seems to be straighforward; if a single, self-
connected event lasts less than one minute it cannot be in a given place at
two times separated by an interval longer than one minute. And the qualifications seem to hold too: in order to be at the same place at two times separated by a one-minute interval, either (i) the event lasts longer than one
minute (and its duration includes the two times) or (ii) the event is composed by two scattered events (as in the case of the successive explosions of
two time bombs that together form one single bombing). Observe that it
appears exactly as meaningful to attribute a duration to scattered bombings as it appears to attribute a length to scattered chairs. In some cases spatially scattered objects form an aggregate whose size is greater than the
sum of the sizes of the places where their parts are located (as it might be in
the case of an army that covers a large battlefield). In some cases spatially
scattered events have a size that can be greater than the sum of the sizes of
the intervals covered by their parts (as it might be in the case of a theater
performance whose two acts are separated by a pause).

Incidentally, that there can be scattered *events* is a question that our
mereotopological theory leaves unsettled, just as it leaves unsettled the
question of whether there can be scattered *objects* in the sense here at
issue. The Fusion Axiom (P.8) guarantees that for every satisfied property
or condition ϕ there is an entity consisting of all those things that satisfy
ϕ. But it does not say that such an entity is itself a ϕ. For every number

of bombings there is a corresponding fusion, but this need not be a bombing—it need not even be an event. For every number of bombs there is a corresponding fusion, but this need not count as a bomb, and some may find reasons not to count it as an object either. Fusions are nothing over and above their constituent parts. But as far as (P.8) is concerned, a (spatial or temporal) fusion of material objects need not be a material object and a (spatial or temporal) fusion of events need not be an event. Once again this is not a matter of stipulation.[9] But as far as structural considerations are concerned, this does not appear to be a source of counterexamples to the complementarity hypothesis: objects and events seem to be on a par in this respect too.

One last point, then. One could think of counterexamples to (O6) by relying on the possibility of time travel. If objects could travel through time, then they *could* be located in different places at the same time. Old John meets young John: they are in front of each other, but they are the same. If this is possible, then (O6) would call for a further qualification to the effect that the object in question has not travelled in time. But then the analogy is obvious. (E6) holds unless the event in question has travelled in space. We have seen reasons to deny that events can indeed travel in space. Shall we conclude that the complementarity hypothesis breaks down because of the possibility of time travel?

Appendix: Extensional Rest and Motion

The various characterizations of movement in section 10.5 involve mereological and locative notions, but they also necessitate that time be represented explicitly: being in motion is being in different places at different times. An alternative account could start from an independent characterization of rest and motion as properties that objects have *at a given time*. An instantaneous snapshot of the world would distinguish between objects that move and objects that are still. This would provide an extensional theory of rest and motion to be added on top of the basic mereotopological and locative framework, as with the theory of dependence of section 5.8 or the theory of holes of section 8.2. In an intuitive conception, both relative and absolute rest and movement are to be taken into account. When one walks inside a moving train, one distinguishes between one's motion relative to the train and one's motion relative to the ground. But of some objects—at least in our unreflective moments—we think that they are not moving, that they are absolutely at rest. The earth is normally not thought of as moving. Very large reference frames, and space itself, are in the norm conceived of as absolutely at rest. We close this chapter with a brief sketch of a theory of common sense rest and motion along these lines.[10]

Let us use a monadic predicate 'Ax', for 'x is at rest absolutely', and a binary predicate 'Axy', for 'x is at rest relative to y'. Using these predicates, we can immediately define corresponding predicates for motion (ignoring for simplicity all fine-grained distinctions between types of motion) :

(10.1) $Mx =_{df} \neg Ax$ (Absolute motion)

(10.2) $Mxy =_{df} \neg Axy$.   (Relative motion)

Relative rest is an equivalence relation, so we fix its basic properties by the following axioms:

(A.1) Axx

(A.2) $Axy \rightarrow Ayx$

(A.3) $Axy \land Ayz \rightarrow Axz$.

The basic properties of relative motion follow as theorems: relative motion is irreflexive and symmetric (though not transitive):

(10.3) $\neg Mxx$

(10.4) $Mxy \rightarrow Myx$.

We then add two axioms to establish a link between absolute and relative rest and motion:

(A.4) $Ax \land Ay \rightarrow Axy$

(A.5) $Ax \land My \rightarrow Mxy$.

Any two objects that are absolutely at rest are at rest relative to each other; and if only one of two objects is at the rest, it moves relative to the other.

This gives us a basic, "Newtonian" picture in which we assume that some objects can be absolutely at rest without explaining this particular property. As an alternative, one could assume the existence of one single individual—call it 'a'—that has this property. (A plausible candidate for a is Newtonian absolute space. In our normal communicative contexts, a is typically the planet Earth.) Given the axiom

(A.6) Aa,

we could then define absolute rest and motion in terms of relative rest and motion with regard to a:

(10.5) A$x =_{df}$ Axa

(10.6) M$x =_{df}$ Mxa.

One could therefore use 'Axy' as the only primitive predicate, now accompanied by a primitive singular term, 'a'. Absolute rest and motion thus defined become *de re* properties, much in the sense in which being John's brother is a property that depends upon John. In the light of such a redefinition, (A.4) and (A.5) are now derivable, since the following are direct consequences of (A.1)–(A.3):

(10.7) A$xa \wedge$ A$ya \rightarrow$ Axy

(10.8) A$xa \wedge$ M$ya \rightarrow$ Mxy.

The formulation in terms of 'a' is thus more powerful. (Of course, instead of choosing as primitives the dyadic predicate 'A' and a constant 'a' for an absolutely resting object, one might also choose 'A' and a constant 'm' for an absolutely moving object, defining 'Mx' as 'Axm', 'Ax' as '$\neg$Mx' and so on.[11] But there are good reasons for not doing so. First, there is only one way of being absolutely at rest, but there are many ways of being absolutely in motion. An object could be moving even though it is not at rest with respect to m, for instance, if they have different speeds or move along different paths. And second, m itself could be moving in many different ways.)

Finally, we could consider integrating the pure theory of rest and motion with the background mereotopological apparatus. For instance, in between the two extremes of relative rest and motion there exist some cases of hybrid motion in which it is not the whole object that moves but only some of its parts (the object is not rigid), or in which the whole object moves but does not change place (as with a disk spinning around its center). In general, these and other intermediate cases of motion occur when different parts of a given object behave differently—with regard to 'A' and 'M'—relative to other objects and their parts. The most general case is *partial motion*:

(10.9) $PMx =_{df} \exists y(Pyx \wedge My)$.

A more specific case is *proper* partial motion, corresponding to the case of an object that has two parts in relative motion:

(10.10) $PPMx =_{df} \exists y \exists z \, (Pyx \wedge Pzx \wedge Myz)$.

More specific cases still will take into account topological distinctions. For instance, *internal* motion obtains when some interior part moves relative to the boundary:

(10.11) $IPMx =_{df} \exists y(Pyx \wedge \forall z(Bzx \wedge Myz))$.

The definition of such predicates is routine. But the exercise may be rewarding. Movement (a kind of event) may provide a good perspective for further testing the degree of independence of spatial thinking from thinking about space—the independence of our way of representing spatial entities from our views (relational or absolutistic) concerning the nature of space.

11 Maps

So far, our study of spatial representation has been mainly a matter of choosing among different theories that differ in descriptive scope, depending on the sort of spatial entities they countenance. But some of these entities are themselves spatial *representations*. Some spatial entities *represent* other spatial entities: they have a semantics. The purpose of this chapter is to lay down the semantic features of the elemental representational spatial items— maps.

Although we believe that the account sketched here could be extended to pictures (photographs, for instance) and other representational spatial items, we think that maps constitute a simpler case for at least two reasons. First of all, maps display no shortenings, no perspective effects: a map can be conceived as close as it gets to a view from nowhere about the area it depicts, whereas a photograph contains additional information about the point of view of the camera. As the tools for spatial representation that we have been developing are meant to capture detached, non-perspectival representations, starting from maps seems therefore most appropriate: maps are as detached as possible. Second, both the map and what it represents can be conceived of as spatial surfaces; and although this constitutes an important methodological simplification, it turns out to be very convenient. (It is not too big a simplification, however. Ideally, our account will be general enough as to include hypothetical voluminous maps of voluminous objects.)

The chapter has three main sections. First, we present a list of data and some hints at the basic problems. Maps are easily used and interpreted, but theoretically not well understood; it is somewhat surprising that there is no basic consensus on what a systematic approach to maps semantics should look like.[1] Second, we introduce the notion of a formal map. A formal map is to an ordinary map what a sentence in a formal language is to a sentence of ordinary language. We then sketch a formal semantics for formal maps based on the mereotopological framework of chapters 3 and 4. The key notions here are those of structured assignment and of color predicate, and the philosophical interesting point is that maps are closer to propositional structures than usually thought.[2] Finally, we offer a survey of problems encountered or expected in extending the semantics to a general project of a semantics for ordinary maps.

11.1 The Data

Maps, as used in everyday life, are fairly simple objects to handle. We produce and interpret maps of various sorts without much trouble. Some of these look quite different from one another, but are all categorised, interpreted, and used as maps. Not much teaching is involved in map reading—much less than is required for learning to speak or read a language. In a way maps are conventional, thereby coming close to linguistic items, more specifically to items of written language. Also, of course, they bear a visual similarity with what they represent, thereby coming close to pictures.

From these data we extract a general characterisation and a program. Maps are representational items and thus have a semantics. They can be used in communication and thus have a pragmatics. They explicitly display information, and further information can inferentially be recovered from them, thus there ought to be a logic that takes maps into account (as premises or conclusions of reasonings, say).

The pragmatic side is not our direct concern here. This in part depends upon the fact that the semantics of an ordinary map is fairly determinate, if compared to the semantics of utterances in ordinary language. Probably the correct term for comparison is a fairly determinate fragment of ordinary language, say, a manufacturer's technical description of a dishwasher. (Very limited uses are possible, contextual effects are minimised, ambiguities are avoided, clarity of expression is a must, technical terms abound, the domain is fixed, and there are only descriptive sentences.) As for the logic side, it turns out to be relatively straightforward (if not trivial) once the semantic aspect is properly understood. Therefore, we shall focus on the general project of a semantics for maps.

By semantics, in the present context, we mean *formal* semantics: a way to systematically associate representational items with semantic values. Existing semantic approaches to maps (e.g., Leong 1994) appear to suffer from the ambition to provide a semantics for maps as these are found and used in everyday life—everyday or *ordinary* maps, as we shall call them in analogy with *ordinary* language.[3] The approach we follow here is different. We try and define the notion of a *formal* map, which is in some sense recognisably similar to that of an ordinary map, and we provide a semantics for (some classes of) formal maps. This is analogous to the project of providing a formal semantics for formal languages. It will then be possible to ask whether some features of this semantics can be used to describe the semantic structure of ordinary maps. The aim is to eventually be in a position to move from the semantics of formal maps to the formal semantics of maps.

11.2 Formal Maps

Maps are spatial objects. They have a shape and a size; they are located somewhere. Spatial structure appears to be an essential property of maps. That maps are themselves essentially spatial helps insofar as spatial features of maps can be conveniently described in the framework of mereotopology. Moreover, maps represent spatial objects, which again can be described in the same mereotopological framework. We shall rely on this fact when formulating semantic conditions for formal maps. As to the formalism, we do not need to go beyond the prospect of ordinary predicate logic, given the general translatability of map encoded information into predicate language.

But why do we need a study of formal, as opposed to ordinary, maps? When we provide a formal semantic account of natural language, we are confronted with problems such as vagueness, ambiguity, semantic closure—phenomena that are resilient to a rigorous coherent account. Formal semantics for natural language turns out to be a nontrivial enterprise. But one can work on the hypothesis that, ideally, to each sentence of natural language there correspond one or more sentences of a suitable formal language that exhibit *logical forms* for the former. And studying these logical forms is a way of studying the semantics of the natural language sentences. Our motivations for studying formal maps are similar. We first investigate the formal semantics for an idealized notion of maps, and then modify it so as to suit the complexity of ordinary maps. This allows us to distinguish between issues of orthography and issues of semantics, and to steer clear of a number of problems such as granularity, ambiguity, and vagueness, which might be raised only once we have firmly established the semantic basis.[4]

In a first approximation, a formal map is a region of space (for ease of discussion, let us assume that a formal map is a nice self-connected, regular square region) associated to a (self-connected, regular) region of the world and variously colored. No names, no lines, no icons are to be found on a formal map: only patches of color. Colors are used here because they provide an unlimited supply of distinct orthographic symbols and are intuitively convenient; one might as well use natural numbers in their stead.

Map subregions need not satisfy any particular shape constraints. For instance, we do not expect them to have any particular shape or size, nor, for that matter, to be self-connected. (Your actual map of Paris might burn except for two disconnected pieces; the mereological sum of these pieces is a map subregion of Paris representing, say, the Bois de Boulogne and Montmartre.)

Color patches are taken to designate properties (of objects in the world). That a certain map region is covered with a certain color patch is taken to mean that the corresponding world region has a certain property (the one corresponding to the color). A uniformly colored map region is correctly colored whenever the corresponding world region has indeed the property corresponding to the color. A non-uniformly colored map region is correctly colored whenever the uniformly colored regions that constitute it are correctly colored. When a map is correctly colored, we may also say that it is *true*. Observe that there are a number of semantic and non semantic features adumbrated here, such as the correspondence of a certain map region with a certain world region, or the location of a color patch within the location of the map.

This sketchy account will be made precise in what follows, but it will also be corrected in some major respect. There is the following problem with the sketchy account. A color signals the presence of a property, but does the absence of a color indicate the absence of that property? Consider the formal map in figure 11.1. Look at the gray region. What are we evaluating when we assign a value to that region? One predicate (gray) or two (gray *and the absence of black*)? And how can we read this off the gray region *only*? But if we cannot read it off a single region, what is the rationale for analyzing a map into regions?

In a sense, determining the semantic value of a map depends on determining the semantic value of its components, so that a certain semantic

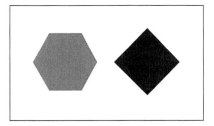

Figure 11.1
A simple formal map.

compositionality is to be expected in the domain of formal maps. Now a main difficulty here is to build up maps recursively out of map components (map regions) in the sense in which one can build up well-formed sentences out of subsentential components (predicative expressions, singular terms, logical constants). Suppose you have a uniformly colored map region: is it composed of its left and right halves or is it composed of its top and bottom halves? What is the next item you semantically evaluate down the compositional scale? On the other hand, the assignment of world regions to map regions is much more structured than the assignment of individuals to singular terms in a first-order language. Once a map region gets assigned a world region, its subregions are associated with subregions of the world region in a tight way. We can somehow expect that the tightness of the assignment compensates for the compositional looseness of maps.

This sketchy account of formal maps indicates that they are propositionally structured—albeit in a peculiar way. A colored map region is like a sentence: it says that a certain individual (a spatial region) has a certain property. Like a sentence, it can be true (if the region is correctly colored) or false (if it is not). Map regions are composed of other map regions, and to some extent colored map regions can be read as conjunctions of map regions. As for the semantic values of map symbols, we can fix them as follows:

(i) *Map regions* are to be considered symbols in themselves, akin to individual constants such as linguistic proper names. A map region has an individual (a world region) as its *only* semantic value.

(ii) *Color patches* are akin to unary predicates. A predicate is an unsaturated expression that becomes saturated by the juxtaposition of an individual constant. A color patch is an unsaturated predicate, which gets saturated when it is juxtaposed to a map region. (We allow ourselves an infinite supply of color patches, of all possible sizes and colors.)

There are no variable symbols among map symbols. This is not to say that maps cannot express generic propositions. However, the generality will not be apparent. Some colors may correspond to predicates containing a hidden quantification, such as 'is hilly', meaning 'there are hills in'. (A predicate such as 'is hilly' is in this respect similar to 'is perforated', which means 'there are holes in'. Both predicates describe properties that hide generality and relational components.) The loss of generality resulting from the absence of variables reduces the inferential scope of maps.

11.3 Map Stages

Once regions of the world have been assigned to map regions, a map can be colored in a number of ways. But we must rule out some of these ways. Suppose the property of being France is represented on a map of Europe by the color red; suppose further that only the southern half of France is colored red, and the northern half remains blank or is colored green. Knowing how things stand, we would surely consider the coloring incomplete and the map false. We can intuitively distinguish between a completely colored false map, and an incompletely colored true map, but it is a fact that we normally assume that a color on a map be distributed up to complete coloring; for each color, the map is not "undercolored" in that color.

Let us then make use of this tacit convention and state it as a formal requirement. Define a *stage* of a given colored map μ to be any coloring of some of regions of μ, with the proviso that a region is colored F in the stage only if it will be so colored in μ. (See figure 11.2 for two examples.) Then a *well-formed stage* of a map μ is a stage of μ such that only one color is present, and it is maximal in μ. More precisely, a well-formed stage of a map μ is a stage σ of μ satisfying the following two conditions:

(11.1)   All colored regions of σ are of the same color.

(11.2)   If a region is painted of that color in μ, it is so painted in σ.

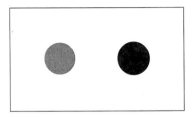

Figure 11.2
Two stages of the map of figure 11.1. The stage on the right is well-formed. The stage on the left is not (both conditions are violated).

As an example, suppose that a map where southern France is painted red is a well-formed stage of a map μ of Europe. Then a map showing southern France painted red and, say, northern Italy painted green is not a well-formed stage of μ by condition (11.1), and a map showing only the southeast quarter of France painted red is not well-formed by condition (11.2).

Reference to well-formed map stages solves at once the problem of compositionality, for it allows us to show how maps can be built out of well-formed map stages. The base case is the empty map. Then there is a set of well-formed atomic map stages with only one color. Then a set of map stages with two colors, and so on until a final well-formed stage with all the colors of the map. (There are no limits to the number of well-formed map stages.) The last stage is, in effect, the map itself.

In a way, this procedure reflects a certain intuition about to how to *print* (rather than draw) a map. The idea is that colors are printed one at a time, with no possibility of going over the same color twice with second thoughts or corrections. It does not matter whether one prints red before green or green before red. When we recursively build a formal map, we just print one color at a time, starting with the empty map. Every coloring of an empty map produces a well-formed map stage, and every new coloring of a map stage (with the exception of a coloring that adopts previously introduced colors) produces a new map stage. Nothing else produces a well-formed map stage.

Well-formed map stages can be evaluated for truth and falsity. They are constructed in a way that leaves no room for indeterminacy relative to a coloring. Requiring that colors are used maximally in a well-formed map stage reflects the fact that the symbols we use to express propositions (colored regions) are intrinsically spatial symbols.

11.4 Formal Semantics for Formal Maps

Having defined our basic concepts, we can now outline a formal seman-
tics for maps. A model for a formal map μ is a structure $M = \langle R, f \rangle$, where
R is a set of world regions and f is an interpretation function such that (i)
for each map regions μ_i of μ, $f(\mu_i)$ is a world region in R; and (ii) for any
pair of (possibly identical) map regions μ_i and μ_j:

(11.3) $P\mu_i\mu_j$ if and only if $Pf(\mu_i)f(\mu_j)$

(11.4) $C\mu_i\mu_j$ if and only if $Cf(\mu_i)f(\mu_j)$.

(An intended model, of course, is one where f assigns to each color the
set of regions that have the property corresponding to the color.) The
assignment is thus structured. When a map region is assigned a world
region, its subregions are thereby assigned subregions of the world region. Starting with the all-inclusive map region, this extends downwards
to all map regions in virtue of clause (i), and is mereotopologically well-
behaved by clause (ii): the mereology and topology of the map regions
mirror the mereology and topology of the corresponding world regions.

The structure of the assignment is expressed in terms of 'P' and 'C' and
is governed by the principles of mereotopology. (We are thinking of
GEMTC, but the account is in principle compatible with any mereotopological theory of the sort discussed in chapter 4.) We need the
mereological constraint in order to avoid situations such as the following:
μ_1 is assigned to France, say, and a region μ_2 that is a proper part of μ_1 is
assigned to the whole of France as well. We need the topological constraint in order to avoid situations such as the following: μ_1 is assigned to
France, and a region μ_2 that is a proper tangential part of μ_1 is assigned
to Burgundy, which is not a tangential proper part of France.

Let us now give the recursive truth conditions for maps. Let σ be an
atomic map stage of μ, i.e., a map stage of μ all colored regions of which
are of the same color, say F. Let σ_1 be the mereological fusion of all such
colored regions, and let σ_2 be the relative complement of σ_1 in σ (i.e.,
$\sigma_2 = \sigma - \sigma_1$). Then σ is true if and only if the following two conditions are
jointly satisfied:

(11.5) $f(\sigma_1)$ has the property associated with color F

(11.6) $f(\sigma_2)$ does not have the property associated with color F.

On this basis, the condition for non-atomic map stages is defined recursively:

(11.7)   A map stage is true if and only if its atomic stages are all true.

A map is just a maximal map stage; hence, (11.7) gives truth conditions for complete maps as well. The truth of a map thus results from the conjoined truth of its well-formed map stages. (The propositionall analogue would be the conjunction of all propositions expressed by each well-formed map stage.[5])

Note that (11.6) plays a crucial condition, which motivates the notion of a well-formed map stage. For if a map does not show a color on a region where it should, it should not be evaluated as true even if every colored region is of the right color. Well-formed map stages solve the problem because they support complementation. (Of course, complementation is one kind of negation. If maps are thought of as sentences, then (11.5) and (11.6) give simultaneous truth conditions for atomic sentences and negations thereof. But we would not be entitled to recursively apply negation to sentences.)

11.4 Ordinary Maps

We have the outline of a formal semantics for formal maps. Let us conclude with some remarks on how this can be extended to a formal semantics for ordinary maps.[6]

First, we must inquire whether there are other *types* of symbols on ordinary maps over and above symbols of the region and color patch type. For instance, are icons and names of localities of the same semantic type as colored regions? Are they merely of different orthographic type (just as the predicates in 'John *is tired*' and 'John *is a geographer*', on an ordinary account), or do they introduce genuinely different semantic ingredients?

Second, we must test the intrinsic expressive limits of color patches. These seem well-suited to express dissective predicates, such as 'part of France' or 'under the direct jurisdiction of the Prince of Liechtenstein'. But color patches are not well-suited for other predicates, such as 'adjacent to Spain' or 'edifiable'.

Third, we must assess the status of redundant symbols. For instance, provided we have enough colors, boundary marks are redundant on maps: they are visually supervenient on the color difference between adjacent regions. Formal maps as described here, in fact, do not explicitly represent boundaries. Taking a different approach, one might want to skip colors altogether and use boundaries to mark the relevant demarcations. Arguably, this calls for a development of the treatment of fiat boundaries briefly broached in chapter 5.

Last, a place should be made for issues of dimensionality, metric, and orientation of a formal map. On the present account, the isomorphism between maps and what they represent is purely mereotopological. As in a subway map, some relevant information is preserved. But, of course, a lot is lost too.

12 Conclusion

We have considered the representation of formal spatial structures under some elementary conditions. First of all, ontological considerations play a methodological role: countenancing different sorts of entities helps to disclose conceptual distinctions that might otherwise be left in the dark. Second, to the extent that spatial representation is concerned directly with these entities (things such as tables and picnics as opposed to purely spatial items such as lines and regions), the structure of these entities must be locked into it; in particular, part-whole structures play a dominant role. Third, an account of these structures involves mereological as well as topological aspects, and neither can be reduced to the other. Fourth, an account of the relation between a spatial entity and its place in space calls for an explicit theory of location: spatial representation goes far beyond the representation of space. These conditions are by no means complete, and we have been silent on a number of related issues. Here, by way of conclusion, is a list of open questions.

A Broader Circle of Spatial Concepts

We concentrated on non-perspectival, detached spatial concepts and representations (such as those associated with 'part of', 'connected to', 'located at'). How do these relate to perspectival ones (such as 'in front of', or 'to the left of')? Do they constitute the inner core of all spatial representation, or are they on an equal footing with the other spatial concepts? In the latter case, is it possible to provide a formal treatment of perspectival concepts so as to make their relations to detached concepts explicit?

Metric Facts

One natural enrichment of mereotopology would take into account metric facts. Although of little salience from a syntactic point of view, metric information is encoded in conceptual representations where it contributes to the understanding of spatial descriptions (witness the difference between 'John crossed the room' and 'John crossed the ocean'). How are metric facts to be accounted for systematically? How do they interact with mereological and topological facts?

Spatial Reasoning

Spatial concepts and representations are used in inferences. The cake is in the tin and the tin is in the kitchen—*hence* the cake is in the kitchen. In this book we have assumed these inferences to be accountable for in a symbolic, propositional format. Psychologically, this would sit well with the hypothesis that one assesses these inferences by using a sort of mental logic. But psychological reality might well be completely different, and an account in terms of mental models might be empirically more adequate. In fact, mental models are generally considered well suited for spatial reasoning. How can the logic of part-whole reasoning be represented in such terms? What is the mental logic of location?

Spatial Logics

If spatial reasoning is symbolic, one still has some options to choose from. The approach followed here treats parthood, connection, and location as relations to be handled in ordinary predicate-logic terms, and theories about these relations take the form of ordinary first-order theories. But a purely propositional-logic account might be viable. For instance, think of parthood as an accessibility relation of the sort usually assumed to hold among possible worlds. (In this case, the "worlds" would be spatial entities and their parts.) What sort of modal logics would correspond to the mereological theories of chapter 3? That is, what modal logics are determined by assuming the accessibility relation to satisfy exactly the mereological axioms of those theories? Some properties of the parthood relation (such as antisymmetry) are not expressible in ordinary modal-logic terms, but richer modal frameworks (with multiple modal operators) might serve the purpose.

Modalities

There are many deep metaphysical questions pertaining to the modal status of the notions that we have studied. We have encountered some of these questions in our discussion of countermereology (chapter 6) and spatial essentialism (chapter 9); however, the intermingling of space and modality is very intricate, and we have tried to stay away from it. To come up with a full-blown picture, one would have to deal with these questions too. One would have to investigate what principles of mereotopology are preserved when parthood and connection are relativized to times and worlds. One would have to see how this extension can combine with a theory of location to provide, for instance, an account of motion in terms of change of location. One would have to deal with counterfactual reasoning about space and spatial reasoning in counterfactual situations.

Absolute vs. Relative Space

On the metaphysical side, we have very cautiously endorsed a moderate form of spatial absolutism—the thesis that space is an independently subsistent individual over and above its inhabitants. This is implicit in our treatment of location, where quantification on independently existing regions of space was assumed from the outset. However, much of what we have been saying could be reinterpreted in relationist terms by replacing each binary predication of the form 'Lxy' with an idiom '$L*x$' that gives a suitable relationist translation. Whether common sense is indeed absolutist or relationist, or whether it has no particular stance on the nature of space, is an open issue. How this in turn may affect the structure of spatial representation is also an open issue.

Naive Topology

We noted several times that common sense has a way of its own to classify objects—in particular, objects with holes (chapter 8)—and to treat topological properties such as contiguity and continuity (chapter 5). In many domains there are remarkable disagreements between common sense and normative topology. More generally, topological classification and understanding of elementary topological facts are impaired by preconceptions about the topological structure of ordinary objects, so that topological equivalences turn out to be counterintuitive to a degree. Some of these preconceptions may involve the dominance of gestalt properties of the visual display of the configuration, the dominance of transformations that preserve metric properties over those that preserve topological properties only, or the assumption that holes are objects of their own. These factors delineate the empirical research field of intuitive topology, a close relative of naive physics.

The Space of Geography

Some of our arguments in this book involved geographic examples. For instance, geography explicitly endorses the distinction between spatial entities and their locations (Italy is not just a boot-shaped territory) as well as the possibility of spatial coincidence of distinct geographic entities (the city of Hamburg and the state of Hamburg). Moreover, geography provides an ideal domain of application for mereotopological concepts and principles (boundaries, for instance, are of central importance). But geography has its puzzles, too. Can a geographic entity survive without a territory? Can it survive radical changes in its territory? Can it exist without definite borders? These questions involve an interesting trade-off between empirical and ontological issues and represent a promising domain of application for the philosophy of spatial representation.

Spatiotemporal (Dis)Analogies

Space has long been the neglected relative in the family of theoretical notions that can be subjected to formal treatment. Time surely has attracted much more interest. One reason for this is that when we introduce a logical treatment of some distinguished notion we happen to start from its encoding in ordinary language. And time is registered in language grammatically in a much more salient way than space. Times and tenses affect the verb, so times can be adequately represented as propositional operators. (Correspondingly, it is facts, not objects, that are future or past.) Space is mostly expressed through prepositions, which are best represented by means of subatomic relational predicates (so that, arguably, it is objects and not facts that are above or inside one another). Do these asymmetries reflect important ontological or cognitive differences between space and time, or are they just linguistic idiosyncrasies?

Spatial Language, Mental Maps, and Localese

We can linguistically describe a spatial display, and we can imagine a spatial display if someone describes it to us. It has long been recognized that at some level there must be an interface between linguistic and spatial representation. One way to bridge the gap between two types of representation would consider the "where" of spatial representation as available in the form of a theory such as the semantic theory for maps sketched in chapter 11, but defined over regions of *mental* maps. Mental map regions would thus be the lexical items—the "locons"—of a spatial quasi-language (a "localese"). One should then provide a characterization of localese that accounts for its hypothetical role of *lingua franca* that translates spatial representations across faculties.

Actual and Possible Concepts

By endorsing a methodological use of common-sense ontology (chapter 1) we did not commit ourselves to any strong notion of common sense. Actual concepts may be significantly different from the concepts of object and event discussed here. (Among other things, we have not taken into account conceptual development or any distinction between infant and adult cognition.) Even so, one might raise the question of the genesis of our concepts by starting from some of the structural elements discussed here. For instance, in relation to the analogies of chapter 10, one might hypothesize that the concept of object and the concept of event are sufficiently similar from a structural point of view to license the hypothesis that they are manifestation of a single, deep underlying concept.

We see here once more how the various senses of spatial representation mentioned in the introduction can interact. Some of the questions we raise are questions of format and scope that pertain to a logic-based, formal approach. Others belong to an *a priori* analysis of the conditions of possibility for certain idealized representational formats that may correspond to actual concepts. In both cases, we think that these questions can be adequately searched only insofar as an articulated formal background exists in which theoretical assumptions are made explicit. For whatever the content of the theory of space one endorses, whatever the logic one subscribes to, whatever the limitations one sets by selecting certain cognitive facts as relevant, at some point the method of philosophy requires—as we may put it—that one calculate.

Notes

Chapter 2

1. The definition below corresponds to the formulation given by Whitehead (1920: 76). Whitehead's earlier definition (1919: 102) is slightly different but essentially equivalent.

2. The figure is adapted from p. 337 of Simons 1987, where this point was first raised; compare also his discussion at pp. 81–86 and Simons 1991d.

3. See *inter alia* Needham 1981.

4. See *inter alia* Bostock 1979.

5. Strictly speaking, a model for Whitehead's theory requires that x and y connect only if they are distinct. In other words, Whitehead's connection relation is irreflexive (whereas de Laguna's and Clarke's are reflexive). The difference turns out to be unimportant. Here we follow Clarke's more convenient interpretation.

6. This definition corresponds to the notion of connectedness in ordinary set-theoretic topology. In this form it was first given philosophical content by Cartwright (1975). Whitehead's formulation was more complex, and Clarke's (1985) formulation involves certain complications that are made necessary by the underlying axiomatic characterization of the 'connection' predicate. (We come back to these complications in chapters 4 and 5.) See also Lewis 1970: 227; Tiles 1981: 56; Hirsh 1982: 97; Van Cleve 1986: 142; Chisholm 1987: 168; Smith 1993: 63.

7. We have discussed the case of holes at some length in chapter 7 of Casati and Varzi 1994. Shadows were suggested by Leibniz in his commentary on Locke (*New Essays*, II-xxvii-1). The ghosts are from van Inwagen 1990: 81. The angels are from Lewis 1991: 75. Chisholm 1973: 590 also mentions shadows and holes in this regard, and Shorter 1977 considers the intriguing case of intersecting clouds produced by two distinct "cloud projectors." The debate on Locke's principle's is still very much alive, and we come back to this point in chapter 7. Here we are simply suggesting that the idea of a one-to-one correspondence between things and places is controversial and would therefore be a bad starting point for a general theory of spatial representation.

8. We come back to the issue of event location in chapter 10.

9. Especially in Casati and Varzi 1994.

10. More discussion of how Flatlanders (from Abbott's 1882 classic) can succeed in determining the shape of their world may be found in the first chapters of Weeks 1985.

11. We come back to this point in chapter 8.

12. The basic idea builds on the work of Marr and Nishihara (1977) and Brooks (1981) and is usually traced back to Binford (1971). Biederman must be given credit for formulating it in purely qualitative terms without resorting to sophisticated abstract hierarchies.

13. We come back to some of these questions at the end of chapter 9.

Chapter 3

1. On the history of mereology see Henry 1991 and Burkhardt and Dufour 1991. For a wealth of annotated references, starting with Brentano, see Smith 1985 and Smith 1993.

2. Actually, the calculus of Leonard and Goodman (1940) had variables for classes. A class-free, purely nominalistic version of the system appeared later (Goodman 1951).

3. There is also another, more controversial sense in which Lewis speaks of the "ontological innocence" of mereology. (That is the sense criticized e.g. in Forrest 1996b.) We come back to this in section 3 below.

4. For thorough and systematic presentations of formal mereologies, see Simons 1987 and Eberle 1970. On Leśniewski's systems, see also Sobociński 1954, Luschei 1965, Clay 1981, and Miéville 1984.

5. A tentative taxonomy may be found in Winston, Chaffin, and Herrmann 1987; see also Iris, Litowitz, and Evens 1988.

6. On this classic position see e.g. Griffin 1977, Wiggins 1980: 300ff, and Thomson 1998.

7. This is the position of Simons (1987), *inter alia*.

8. For more on the opposition between parts and components see Sharvy 1983a and Sharvy 1983b. On the mereology of artifacts, see Hexmoore 1990 and Simons and Dement 1996.

9. See Lyons 1977: 313, Cruse 1979, Winston, Chaffin, and Herrmann 1987, Iris, Litowitz, and Evens 1988, Sanford 1993, Gerstl and Pribbenow 1995, Moltmann 1997, *inter alia*.

10. This is the typical example discussed by most of the authors mentioned in the previous note.

11. Many formal mereologies have been developed using other primitives as well, e.g., the relation of overlap defined below. There is no substantial difference between these various options.

12. See for instance Thomson 1983, Simons 1987, and Simons 1991b.

13. See Aczel 1988 for a general presentation, and Barwise and Etchemendy 1987 and Barwise and Moss 1996 for applications and developments.

14. This choice of logic is not without consequences. Particularly when it comes to the mereological operators of sum, product, etc., a free logic would arguably provide a more adequate background (see Simons 1991b). However, our purpose here is modest enough to justify the simplifications afforded by the assumption of a classical logic (with descriptive terms treated *à la* Russell (1905)).

15. Here and in the following we simplify notation by dropping all initial universal quantifiers. All formulas are therefore to be understood as universally closed.

16. A thorough discussion can be found in Eberle 1970 and Simons 1987.

17. See Simons 1991b as well as Baumgartner and Simons 1994 and Simons and Dement 1996. The term 'supplementation' is also due to Simons.

18. This reading of Brentano is defended by Chisholm (1978). For an assessment, see Baumgartner and Simons 1994. Of course one can simply reject Brentano's premise here: John and thinking-John are one and the same.

19. Early formulations of the objection may be found in Hempel 1953 and Rescher 1955. See also Eberle 1970: 81-99 for an original attempt to accommodate the idea of "sequential individuals" within an extensional framework.

20. Withdrawals of mereological extensionality based on this sort of argument may be found in Wiggins 1979, Doepke 1982, Simons 1987, and Lowe 1989. Some authors, most

notably Chisholm (1973), prefer taking the puzzle as a *reductio ad absurdum* of the idea that an object can undergo mereological change, while others, most notably van Inwagen (1981), take it to be a *reductio* of unrestricted mereological realism, i.e., the idea that undetached bodily parts (such as the tail or the rest of the body before the annihilation of the tail) are actual entities. We come back to some of these options in chapters 8 and 6, respectively. Of course, one remaining option would be to bite the bullet and give up transitive identity: see e.g. Garrett 1985. (Geach (1967) and Noonan (1980) take identity to be relative to sortal terms, with similar results.)

21. Our argument here is similar to that of Della Rocca (1996), though our concern is exclusively with mereological sums rather than material constitution. In a similar fashion, Neale (1990, section 4.6) argues that this sort of *de re* / *de dicto* ambiguity is responsible for a certain confusion surrounding the discussion on events and event identity. (See e.g. the reasoning in Brand 1977: 334.) All these arguments owe much to Smullyan (1948) and Kripke (1972). (See the discussion of the identity of heat and molecular motion in Kripke 1972: 129ff.)

22. Such an account is sketched e.g. in Thomson 1983. It would seem that the only way to avoid this modal step (and keep transitive identity along with mereological extensionality, mereological realism, and the possibility of mereological change) would be to endorse a four-dimensional ontology, as in Heller 1984 or Sider 1997. The cat and the sum of her body and her tail would then be distinguishable on account of their having different temporal extent, hence different temporal parts.

23. This is the obvious modal extension of Chisholm's (1973) idea that ordinary objects are *entia successiva*.

24. We are assuming a classical logical background, treating 'ι' as contextually defined *à la* Russell. Much of what we say, however, would also apply in case we used a free logic instead, with 'ι' assumed as part of the logical vocabulary proper.

25. This elaborates on an example of Baxter's (1988a: 579). Other authors of course would emphasize that the number of objects cannot be established except with reference to some conceptual scheme. See e.g. Putnam 1987, Part II.

26. The classical criticisms are in Lowe 1953 and again Rescher 1955, with replies in Goodman 1956 and Goodman 1958. Other criticisms include Chisholm 1973, Wiggins 1980, van Inwagen 1987, and van Inwagen 1990. See van Cleve 1986, Sider 1997 (section 3.1), and Rea 1998 for articulated replies.

27. The psychological inclination to do away with gerrymandered sums has also a linguistic side: we typically don't have terms for such entities. Even Russell's notion of a logically proper name reflects this inclination: "from a logical point of view, a proper name may be assigned to any continuous portion of space-time." (1940: 33) But we can only agree with Chomsky that this view embodies "an empirical hypothesis. Interpreted in this way Russell is stating what is, no doubt, a psychological truth. Interpreted otherwise, he is giving an unmotivated definition of 'proper name'. There is no logical necessity for names or other 'object words' to meet any condition of spatiotemporal contiguity or to have other Gestalt qualities, and it is a nontrivial fact that they apparently do, insofar as the designated objects are of the type that can actually be perceived (for example, it is not true of 'United States'—similarly, it need not be true of somewhat more abstract and functionally defined notions such as 'barrier')." (Chomsky 1967: 29)

28. See Martin 1965 and Bunt 1985; see also Bunge 1966 and Bunge 1977 for a theory with *several* null individuals.

29. See Tarski 1935. Compare also Clay 1974 for an exact assessment of this result in relation to Leśniewski's systems, which are not based on classical logic.

30. For more discussion on atomism and related issues, see Eberle 1970 and references therein.

Chapter 4

1. Whitehead's connection relation is irreflexive, but this is "merely a convenient arrangement of nomenclature." (Whitehead 1929: 295) See Gerla and Tortora 1992 for a reformulation of Whitehead's system based on a reflexive notion of connection. (Clarke's 1981 reformulation is also based on reflexivity, but the resulting system is not equivalent to Whitehead's in a number of respects.)

2. See Cohn and Varzi 1998 for a comparative study of some of these options.

3. See Varzi 1994 for more preliminary material on this taxonomy of strategies.

4. We used a mereotopology of this sort in the appendix of Casati and Varzi 1994.

5. Similar theories have been proposed by various other authors, though often with a different choice of primitives and/or axiom sets. For instance, Smith's (1993) version uses 'IP' as a primitive; Pianesi and Varzi (1994a, 1994b, 1996b) give an equivalent formulation using as a primitive the closure operator 'c' defined below. A version of the same theory based on a different 'boundary' primitive is found in Pianesi and Varzi 1996a. A slightly weaker formulation, based on 'IP', can also be found in Tiles 1981. It should also be mentioned that the strategy of relying on the intuitive notion of an object being *wholly within* another goes as far back as Nicod (1924). Lejewski's (1982) idea of using the relation of an interval being *wholly earlier* than another (see section 4.1 above) exploits the same intuition and so does much linguistics-oriented work on time, tense, and aspect: see *inter alia* Kamp 1979, van Benthem 1983, Bach 1986, Link 1987, Krifka 1989, Landman 1991.

6. The correspondence is obtained by interpreting 'P' as set inclusion and '+' as union. Likewise, in (C.7′) below '×' is the mereotopological analogue of set intersection.

7. Some authors, including Grzegorczyk (1960) and Kamp (1979), have gone as far as construing boundaries as higher-order entities, but this is somewhat paradoxical if the theory allows for boundary individuals. By contrast, the necessity of countenancing boundaries has been stressed by Smith (1993). We postpone a more detailed account of the literature to chapter 5.

8. Biacino and Gerla (1991) call such systems 'Connection Structures' (we are obviously using this term in a more general sense). See Gerla 1995 for an overview of the relevant literature.

9. This way of proceeding is considered in Varzi 1994.

10. See also Vieu 1991 and Aurnague and Vieu 1993b. A complete axiomatization can be found in Asher and Vieu 1995.

11. Compare Randell 1991; Randell and Cohn 1992; Randell, Cui, and Cohn 1992b; Cohn, Randell, and Cui 1995. For developments, see also Gotts 1994a; Gotts 1994b; Cohn and Gotts 1994; Gotts, Gooday, and Cohn 1996; Cohn, Bennett, Gooday, and Gotts 1997. Some discussion of this theory will be found in the next chapter.

12. The restrictions in (4.37) can be dropped: C is unrestrictedly symmetric.

Chapter 5

1. This issue has been addressed in the literature both with regard to its cognitive and ontological underpinnings and, perhaps more indirectly, in the context of related research in natural language semantics and spatiotemporal reasoning. Some indicative examples are Stroll's work on surfaces (1979, 1985, 1988), Adams's operationalist topology (1973, 1984, 1986, 1996), Gibson's (1966, 1979) ecological psychology, and Jackendoff's (1991) work on conceptual semantics along with more specific work such as Chisholm's (1984, 1992/93,

1994) and Smith's (1993, 1995a, 1995b, 1996) investigations into Brentano's theory of the continuum. On the historical background of the debate, see Zimmerman 1996a.

2. Here, as elsewhere in the book, we cautiously endorse a form of methodological realism: we treat boundaries as objects of inquiry insofar as they play a role in the representation of spatial facts. If our position is read in the frame of a debate on the reality of middle-size objects, then its contribution is only to the descriptive part of the debate. If one is a realist about boundaries, then one must find in the world a place for entities with the properties here described. If one is an irrealist, then one must find weapons to explain boundaries away, in the face of their contribution to folk psychological explanations.

3. The method was first presented in Whitehead 1916, but it is in Whitehead 1929 that we find the most refined formulation of the theory. The approach was fully worked out in Clarke 1985 (see also Biacino and Gerla 1996) and it is this formulation that we consider here. One should also mention here, in the temporal realm, the works of Hamblin (1969, 1971) and Allen (1981, 1984).

4. Some authors, most notably Stroll (1988), have argued that every boundary can be viewed either as an abstraction or as a physical feature, and that this conceptual vacillation cannot be resolved. We do not pursue this hypothesis here, though we consider a way of dealing with the opposition in section 5.7.

5. Besides de Laguna's work, one should mention here the constructions of Broad 1923, Nicod 1924, Russell 1927, Tarski 1929, and Menger 1940. See also Gerla 1995 for an overview and Forrest 1996a for a variant. The abstractive approach has an analogue in the temporal realm, where instants are sometimes construed as sets of time intervals (which in turn are sometimes construed as sets of overlapping events). The *locus classicus* is Russell's (1914) construction, echoed in Walker 1947 and Kamp 1979. For a thorough philosophical defense of the approach, see Hazen 1990, 1993.

6. This would echo von Wright's (1963) treatment of events as couples of states (intuitively, the state obtaining before the event and the state resulting from the occurrence of the event).

7. Randell and Cohn (1989: 361), building on Clarke's theory, actually adopt a version of (5.1b). (This is rejected in later works: see e.g. Randell, Cui, and Cohn 1992b: 173.)

8. To be sure, the exact status of undefined terms in the theory depends on the underlying semantics for the definite description operator 'ι'. One might want to say that even though (C.9) implies that no boundaries *exist*, nonetheless (5.1a)–(5.1d) provide a definition of the *concept*, and therefore the system is not entirely boundary-free. This is an interesting issue, but addressing it here would take us too far afield.

9. See there, note 11 for bibliographic references.

10. It is apparent from the quote that the authors take their theory to be concerned with relations among *regions* rather than objects. But their motivations would apply equally well if we take mereotopology to apply to full-fledged spatial entities.

11. This is the strategy of Smith 1993.

12. These equivalence stems from Zarycki 1927. As always, we are assuming the identities in (5.17)–(5.19) to hold whenever b is defined for its arguments.

13. On the idea that material objects have open complements, see also Asher and Vieu 1995. Richard Cartwright (1975) has argued that all bodies (more precisely, all receptacles of material bodies) are open, their surface always being *between* them and their comple- ment. (This reflects Descartes' notion of "superficies": see *Principles of Philosophy*, Part 2, Principle 15.) Of course *this* option—all closed vs. all open—is "hardly worth of serious dispute," as Cartwright says (p. 154). Our point here is that the option should rather include the possibility of some things being closed and some others being open. On these issues, see also Helen Cartwright's (1987) discussion.

14. For more on this interpretation see Desclés 1989, Jackendoff 1991, Pianesi and Varzi 1994a, Piñón 1997, and Giorgi and Pianesi 1997 (chapter 4).

15. As we have seen, the two halves may each include half of the boundary and thereby qualify as partly closed and partly open. Henceforth we ignore the complications that arise from such considerations.

16. The image is from Adams 1984: 400. In the same spirit, Zimmerman (1996b: 25) speaks of "a part of the object which was first imbedded in the [object] and then disclosed wearing a new skin."

17. Further examples are described in Casati and Varzi 1994 (chapter 10) and Galton 1999 (chapter 6).

18. For instance along the lines of Davis 1993.

19. Compare Kline and Matheson 1987 for a related argument. A Whiteheadian would not have this problem, as clarified in Hazen 1990. Further discussion in Godfrey 1990.

20. We endorse Smith's term 'fiat boundary' also in order to avoid the problems mentioned in section 4.2 with regard to the term 'abstraction'. The opposition between fiat and salient (bona fide) boundaries is further investigated in Smith and Varzi 1999a.

21. Zimmerman (1996a) made us realize that this view seems to have an ancestor in Suarez (*Disputationes*, § 56). See also Zimmerman 1996b.

22. This draws on Brentano's view, which regards the possibility of coincidence as a distinguishing feature of all boundaries. (See e.g. Brentano 1976: 41 and Brentano 1924: 357f. Detailed formulations have been worked out by Chisholm (1984, 1992/1993) and Smith (1995a, 1997b).) However we are not suggesting to embrace Brentano's as a general theory of boundaries. It is when it comes to fiat boundaries that coincidence may enter the picture. See Smith and Varzi 1999a for a detailed account of this second option.

23. Authors in the boundary-free tradition of Whitehead and Clarke have a corresponding problem, for we may directly speak of vague *objects* or *regions*. See Cohn and Gotts 1996 for some work in this area. See also the papers collected in Burrough and Frank 1996.

24. The idea of a vague or "rough" mereology has been given formal treatment in Polkowsky and Skowron 1994, where parthood undergoes a fuzzification parallel in many ways to the fuzzification of membership in Zadeh's (1965) fuzzy set theory. On vague parthood see also van Inwagen 1990 (chapter 13), Sylvan and Hyde 1993, and Copeland 1995.

25. The view that vague boundaries are to be explained away in semantic terms has been defended in very similar terms by Quine (1985). See also Mehlberg 1956 (section 29), for an early formulation.

26. See Brentano 1976, Part I (and compare Aristotle, *Metaphysics*, K 1060b12ff). The ontological dependence of boundaries has been stressed by Chisholm (1984, 1994), Hestevold (1986), Bochman (1990), and especially Smith (1993). Compare also Stroll 1988 (chapter 2), on the dependence of surfaces.

27. On ontological dependence see, e.g., Simons 1987 (chapter 7), Johansson 1989 (chapter 9), and Fine 1995b.

Chapter 6

1. Some empirical evidence is reported in Kosslyn, Heldmeyer, and Glass 1980, in Tversky and Hemenway 1984, and in Tversky 1989.

2. We have toyed informally with potential parts in Casati and Varzi 1994 (chapter 2). Here we correct that line of thought.

3. See chapters 5 and 7 of Simons 1987 for a review of the main problems, some of which occupy us in chapter 8 below.

4. This would be in the spirit of Thomson 1983, with worlds in place of times.

5. In the third *Logical Investigation* (1900/1901), especially in sections 2–5 and 13. The distinction actually goes back at least to Stumpf (1873), which is where Husserl got it.

6. In section 5.8 we did not go into the question of whether a boundary is rigidly dependent on its host (i.e., can only exist as a boundary of that host) or only generically dependent (i.e., can only exist as a boundary of an object such as its host). However, this is irrelevant here. For even if we assume boundaries to be only generically dependent on their hosts, their dependence prevents them from ever existing in isolation, hence from ever being *in a world* in the sense relevant here. (See Smith 1992, section 10, for further discussion of the generic dependence of boundaries upon their hosts.)

7. Some complications arising from this view are discussed in Carter 1983.

8. If we think of the world as the mereological sum of everything in it, then (M) amounts to the requirement that any inventory of the world be a list of things of which the world is *strictly made up*, in the sense of Chisholm (1973: 587), or of which the world is *composed*, in the sense of van Inwagen (1987: 22).

9. On the mereotopology of events and the corresponding variety of admissible count policies, see Pianesi and Varzi 1996a and Pianesi and Varzi 1996b, respectively.

Chapter 7

1. Vagueness issues do not arise here even if one rejects our conception of vagueness as a merely semantic phenomenon (see section 5.8): if Mount Everest had genuinely fuzzy boundaries, it would still have an exact location (though the boundaries of this location would be correspondingly fuzzy).

2. See Perzanowski (1993) for a different axiomatic treatment of locative relations.

3. Thus, in particular, 'R' here is different from the region predicate 'R' mentioned in section 4.5 in relation to Eschenbach and Heydrich's (1995) work.

4. See Randell, Cui, and Cohn (1992b: 173) for a similar remark about **SMT**-type mereo-topologies.

5. Actually, the treatment of Brentano's notion of coincidence put forward by Smith (1997b) is based on postulates closely resembling the properties of RL just outlined.

6. Locke's principle has been discussed *inter alia* by Shorter (1977), Burke (1980, 1992, 1994, 1997), Doepke (1982, 1986), Simons (1985, 1986), Noonan (1986), Lowe (1995), Oder-berg (1996), Olson (1996), Levey (1997), Rea (1997), Zimmerman (1997), Thomson (1998). Much of the debate has originated with Wiggins 1968. See also Sorabji 1988 for a critical review of pre-Lockean discussions of the principle of one object to a place.

Chapter 8

1. See for instance Cohn, Randell, and Cui 1995. For similar accounts in terms of convex locative inclusion, see Wunderlich 1985, Habel 1989, and Herweg 1989.

2. For more on this, see chapter 5 of Casati and Varzi 1994.

3. The reader may find this material in the appendix to Casati and Varzi 1994. The axiomatization of 'H' is further developed in Varzi 1996a.

4. For more on this distinction, see chapter 4 of Casati and Varzi 1994. Note that the theory of boundaries allows us to give a complete account of the distinction without introducing morphological notions (such as the notion of filling) or further topological concepts (such as an operator to account for differences in topological genus) as we did in our 1994 work. It is also instructive to compare these definitions with those of Gotts (1994a, 1994b).

5. Entry boundaries correspond to free faces in the terminology of our 1994 work.

Chapter 9

1. See Noonan 1993 for a collection of texts.

2. See e.g. Plantinga 1975, Wiggins 1979, Van Cleve 1986, Willard 1994.

3. Some authors would add the requirement that x must also exist in those worlds in which y exists and x is part of y. We shall not consider this part of the story and refer the reader to chapter 7 of Simons 1987 for discussion. Likewise, we shall ignore here the variants of mereological essentialism that can be obtained by making parthood interact with time or with other parameters, as initially suggested by Plantinga (1975).

4. At least this is true if the connection relation is assumed to satisfy certain plausible conditions that we have examined in chapter 5. We have seen in section 5.5 that Clarke's (1981) mereotopology does not obey this principle.

5. A formulation of this sort of thinking can be found in Denkel 1995 *contra* Burke 1994. See also van Inwagen 1981.

6. In the terminology of chapter 8, we could define 'D' to be a predicate that is true of an object x iff x has a perforating hole. Then (DD) would be analytic.

7. Here van Inwagen is talking about a principle he calls 'positional essentialism', which effectively corresponds to our $\mathcal{R}$IN below.

Chapter 10

1. Other philosophers who propounded a four-dimensional conception of objects, without necessarily identifying them with events, include Smart (1972, 1982), Noonan (1976), Heller (1984), McCall (1994), and Sider (1997). Echoes may also be found in some AI literature, as with Hayes's (1985a) notion of a "history." For further (annotated) references to the extensive literature on events, see Casati and Varzi 1997a.

2. In the sense of Strawson 1959.

3. Elsewhere, Wiggins uses terms such as 'dual' or 'transposition' to describe phenomena that we dub 'complementary'. Views similar to Wiggins's have been put forward *inter alia* by Mellor (1980, 1995) and Brand (1982, 1984, 1989).

4. We are thinking, especially, of the early debate among Taylor (1955, 1959), Mayo (1961), Dretske (1962, 1967), and Thomson (1965). The philosophical literature on time travel is now very extensive (see e.g. Ray 1991), but here we are only concerned in the topic insofar as it pertains to the issue of spatio-temporal analogies.

5. Mayo uses 'occupy' instead of 'take up', but with no suggestion that occupation amounts to exclusive location.

6. See section 7.7 on the connection between location and occupation.

7. For an attenuation, see Armstrong 1980 and Robinson 1982.

8. Some limit cases are less clear. Take a disk that neither spins nor translates, but simply

expands and contracts. Here it is unclear whether the concept of motion applies at all. See Forrest 1984 and Sayan 1996 for further cases of dubious motion in the sense of (D').

9. The strong form of (P.8), according to which events are closed under fusion, is endorsed e.g. by Thomson (1977). See B. Taylor 1985: 25 and Lombard 1986: 25 for misgivings. (Lewis is not so liberal either: see Lewis 1986.) Compare also the discussion in chapter 10 of Bennett 1988.

10. A more detailed discussion is in Casati 1998.

11. Choosing as primitives both 'm' and the dyadic 'M' would not do. One would not be able to define 'Mx' as 'Mxm', for x could be at rest.

Chapter 11

1. A great deal of semiotic research on maps has concentrated on the classification of the types of symbols one may find on existing maps, and on some normative directions as to the admissible types of symbols. For two proposals in the spirit of formal semantics, see Pratt 1993 and Leong 1994.

2. One interesting side question concerns the linguistic/pictorial interface. The use of maps is sometimes related to questions from users ('Is Manhattan south of Rome?') as related to information stored in a map, and to the necessity of translating this information back into answers to the users in linguistic format ('Manhattan is south of Rome').

3. We do not wish to delve into the issue of whether there are *natural* mental maps in the sense in which there are *natural* languages, so we stick to the qualifier 'ordinary'.

4. In section 11.4 we mention the possibility of reducing all map symbols to two semantic categories paralleling the familiar categories of *individual constants* and *predicates*.

5. To evaluate the truth of map regions that are proper parts of a map we can imagine that a similar procedure applies. A *map region stage* will be defined relative to each color, and conditions such as (11.5) and (11.6) will apply. We would still need well-formed map stages, as opposed to map region stages, to ensure compositionality.

6. For a more extended discussion see Casati 1999b.

References

In the text and footnotes, bibliographic references are given by name and year of publication of the original edition. In the case of quoted texts, page numbers refer to the latest edition (or translation) indicated here. We also list here titles that are not explicitly referred to in the text, but which have nonetheless inspired some of our remarks. By contrast, works by classic philosophers such as Aristotle, Locke, and Leibniz are not listed, but references are given in a way that allows the use of most editions. The quotations from Aristotle's Metaphysics *are from the translation by Richard Hope (New York: Columbia University Press, 1952).*

Abbott E. A., 1882, *Flatland, A Romance of Many Dimensions*, London: Seeley (reprinted by Penguin Books, 1952).

Aczel P., 1988, *Non-Well-Founded Sets*, Stanford: CSLI Publications.

Adams E. W., 1973, 'The Naive Conception of the Topology of a Surface of a Body', in P. Suppes (ed.), *Space, Time and Geometry*, Dordrecht: Reidel, 402–424.

Adams E. W., 1978, 'Two Aspects of Physical Identity', *Philosophical Studies* 34, 111–134.

Adams E. W., 1984, 'On the Superficial', *Pacific Philosophical Quarterly* 65, 386–407.

Adams E. W., 1986, 'On the Dimensionality of Surfaces, Solids, and Spaces', *Erkenntnis* 24, 137–201.

Adams E. W., 1996, 'Topology, Empiricism, and Operationalism', *The Monist* 79, 1–20.

Allen J. F., 1981, 'An Interval-Based Representation of Temporal Knowledge', *Proceedings of the 7th International Joint Conference on Artificial Intelligence*, San Mateo (CA): Morgan Kaufmann, 221–226.

Allen J. F., 1984, 'Towards a General Theory of Action and Time', *Artificial Intelligence* 23, 123–154.

Armstrong D. M., 1968, *A Materialist Theory of the Mind*, London: Routledge & Kegan Paul.

Armstrong D. M., 1980, 'Identity through Time', in P. van Inwagen (ed.), *Time and Cause: Essays Presented to Richard Taylor*, Dordrecht: Reidel, 67–78.

Asher N. and Vieu L., 1995, 'Toward a Geometry of Common Sense: A Semantics and a Complete Axiomatization of Mereotopology', in *Proceedings of the 14th International Joint Conference on Artificial Intelligence*, San Mateo (CA): Morgan Kaufmann, 846–852.

Aurnague M. and Vieu L., 1993a, 'A Three-Level Approach to the Semantics of Space', in C. Z. Wibbelt (ed.), *The Semantics of Prepositions: From Mental Processing to Natural Language Processing*, Berlin: de Gruyter, 393–439.

Aurnague M. and Vieu L., 1993b, 'Toward a Formal Representation of Space in Language: A Commonsense Reasoning Approach', in F. Anger, H. Guesgen and J. van Benthem (eds.), *Proceedings of the Workshop on Spatial and Temporal Reasoning. 13th International Joint Conference on Artificial Intelligence*, Chambéry: IJCAI, 123–158.

Bach E., 1986, 'The Algebra of Events', *Linguistics and Philosophy* 9, 5–16.

Bäckström C., 1990, 'Logical Modelling of Simplified Geometrical Objects and Mechanical

Assembly Processes', in Su-shing Chen (ed.), *Advances in Spatial Reasoning, Volume 1*, Norwood: Ablex, 35–61.

Barwise J. and Etchemendy J., 1987, *The Liar: An Essay in Truth and Circularity*, Oxford: Oxford University Press.

Barwise J. and Moss L., 1996, *Vicious Circles: On the Mathematics of Non-Wellfounded Phenomena*, Stanford: CSLI Publications.

Baumgartner W. and Simons P., 1994, 'Brentano's Mereology', *Axiomathes* 5:1, 55–76.

Baxter D., 1988a, 'Identity in the Loose and Popular Sense', *Mind* 97, 575–582.

Baxter D., 1988b, 'Many-One Identity', *Philosophical Papers* 17, 193–216.

Bennett B., 1994, 'Spatial Reasoning With Propositional Logics', in J. Doyle, E. Sandewall, and P. Torasso (eds.), *Principles of Knowledge Representation and Reasoning: Proceedings of the Fourth International Conference*, San Mateo (CA): Morgan Kaufmann, 51–62.

Bennett J., 1988, *Events and Their Names*, Indianapolis: Hackett.

Biacino L. and Gerla G., 1991, 'Connection Structures', *Notre Dame Journal of Formal Logic* 32, 242–247.

Biacino L. and Gerla G., 1996, 'Connection Structures: Grzegorczyk's and Whitehead's Definitions of Point', *Notre Dame Journal of Formal Logic* 37, 431–439.

Biederman I., 1987, 'Recognition-by-Components. A Theory of Human Image Understanding', *Psychological Review* 94, 115–147.

Biederman I., 1990, 'Higher-Level Vision', D. N. Osherson, S. M. Kosslyn, and J. M. Hollerbach (eds.), *An Invitation to Cognitive Science. Volume 2: Visual Cognition and Action*, Cambridge (MA): MIT Press, 1–36.

Binford T. O., 1971, 'Visual Perception by Computer', paper presented at the IEEE Systems Science and Cybernetics Conference, Miami, December.

Bloom P., Peterson M. A., Nadel L., and Garrett M. F. (eds.) 1997, *Language and Space*, Cambridge (MA): MIT Press (Bradford Books).

Bochman A., 1990, 'Mereology as a Theory of Part-Whole', *Logique et Analyse* 129/130, 75–101.

Bolzano B., 1851, *Paradoxien des Unendlichen*, ed. F. Přihonský, Leipzig: Reclam (Eng. trans. by D. A. Steele, *Paradoxes of the Infinite*, London: Routledge & Kegan Paul, 1950).

Borges J. L., 1949, 'El Aleph', in *El Aleph*, Buenos Aires: Losada (Eng. trans. by A. Kerrigan, 'The Aleph', in *A Personal Anthology*, New York: Grove, 1967).

Borgo S., Guarino N., and Masolo C., 1995, 'A Naive Theory of Space and Matter', in P. Amsili, M. Borillo, and L. Vieu (eds.), *Time, Space and Movement: Meaning and Knowledge in the Sensible World. Proceedings of the 5th International Workshop*, Toulouse: COREP, Part E, 29–32.

Borgo S., Guarino N., Masolo C., 1996, 'A Pointless Theory of Space Based on Strong Connection and Congruence', in L. Aiello, J. Doyle, and S. C. Shapiro (eds.), *Principles of Knowledge Representation and Reasoning: Proceedings of the Fifth International Conference*, San Francisco (CA): Morgan Kaufmann, 220–229.

Bostock D, 1979, *Logic and Arithmetic, Volume 2: Rational and Irrational Numbers*, Oxford: Clarendon.

Braddon-Mitchell D. and Jackson F., 1997, *Philosophy of Mind and Cognition*, Oxford: Blackwell.

Brand M., 1977, 'Identity Conditions for Events', *American Philosophical Quarterly* 14, 329–337.

Brand M., 1982, 'Physical Objects and Events', in W. Leinfellner, E. Kraemer, and J. Schank (eds.), *Language and Ontology. Proceedings of the 6th International Wittgenstein Symposium*, Vienna: Hölder-Pichler-Tempsky, 106–116.

Brand M., 1984, *Intending and Acting. Toward a Naturalized Action Theory*, Cambridge (MA): MIT Press (Bradford Books).

Brand M., 1989, 'Events as Spatiotemporal Particulars: A Defense', in W. L. Gombocz, H. Rutte, and W. Sauer (eds.), *Traditionen und Perspektiven der analytischen Philosophie. Festschrift für Rudolf Haller*, Vienna, Hölder-Pichler-Tempsky, 398–414.

Brentano F., 1924, *Psychologie vom empirischen Standpunkt, 2. Aufgabe*, ed. O. Kraus, Leipzig: Meiner (Eng. trans. ed. by L. L. McAlister, *Psychology from an Empirical Standpoint*, London: Routledge & Kegan Paul, 1950).

Brentano F., 1933, *Kategorienlehre*, ed. A. Kastil, Hamburg: Meiner (Eng. trans. by R. M. Chisholm and N. Guterman, *The Theory of Categories*, The Hague: Nijhoff, 1981).

Brentano F., 1976, *Philosophische Untersuchungen zu Raum, Zeit und Kontinuum*, ed. S. Körner and R. M. Chisholm, Hamburg: Meiner (Eng. trans. by B. Smith, *Philosophical Investigations on Space, Time and the Continuum*, London: Croom Helm, 1988).

Broad C. D., 1923, *Scientific Thought*, New York: Harcourt.

Brooks R., 1981, 'Symbolic Reasoning among 3-D Models and 2-D Images', *Artificial Intelligence* 17, 285–348.

Bunge M., 1966, 'On Null Individuals', *Journal of Philosophy* 63, 776–778.

Bunge M., 1977, *Treatise on Basic Philosophy, Volume 3. Ontology I: The Furniture of the World*, Dordrecht: Reidel.

Burge T., 1977, 'A Theory of Aggregates', *Noûs* 11, 97–117.

Bunt H. C., 1985, *Mass Terms and Model-Theoretic Semantics*, Cambridge: Cambridge University Press.

Burkhardt H. and Dufour C. A., 1991, 'Part/Whole I: History', in H. Burkhardt and B. Smith (eds.), *Handbook of Metaphysics and Ontology*, Munich: Philosophia, 663–673.

Burke M. B., 1980, 'Cohabitation, Stuff and Intermittent Existence', *Mind* 89, 391–405.

Burke M. B., 1992, 'Copper Statues and Pieces of Copper: A Challenge to the Standard Account', *Analysis* 52, 12–17.

Burke M. B., 1994, 'Preserving the Principle of One Object to a Place: A Novel Account of the Relations Among Objects, Sorts, Sortals, and Persistence Conditions', *Philosophy and Phenomenological Research* 54, 691–624.

Burke M. B., 1994, 'Dion and Theon: An Essentialist Solution to an Ancient Puzzle', *Journal of Philosophy* 91, 129–139.

Burke M. B., 1996, 'Tibbles the Cat: A Modern *Sophisma*', *Philosophical Studies* 84, 63–74.

Burke M. B., 1997, 'Coinciding Objects: Reply to Lowe and Denkel', *Analysis* 57, 11–18.

Burrough P. A. and Frank A. U. (eds.), 1996, *Geographic Objects with Indeterminate Boundaries*, London: Taylor and Francis.

Campbell J., 1993, 'The Role of Physical Objects in Spatial Thinking', in N. Eilan, R. McCarthy, and B. Brewer (eds.), *Spatial Representation. Problems in Philosophy and Psychology*, Oxford: Blackwell, 65–95.

Campbell J., 1994, *Past, Space and Self*, Cambridge (MA): MIT Press (Bradford Books).

Carey S., 1988, 'Conceptual Differences Between Children and Adults', *Mind and Language* 3, 167–181.

Carey S., 1991, 'Knowledge Acquisition: Enrichment or Conceptual Change?', in S. Carey and R. Gelman (eds.), *The Epigenesis of Mind*, Hillsdale (NJ): Erlbaum, 257–291.

Carey S., 1993, 'Speaking of Objects, as Such', in G. Harman (ed.), *Conceptions of the Human Mind*, Hillsdale (NJ): Erlbaum, 139–159.

Carey S. and Xu F., 'Infants' Metaphysics: The Case of Numerical Identity', *Cognitive Psychology* 30, 111–153.

Carter W., 1983, 'In Defense of Undetached Parts', *Pacific Philosophical Quarterly* 64, 126–143.

Cartwright H. M., 1987, 'Parts and Places', in J. J. Thomson (ed.), *On Being and Saying: Essays for Richard Cartwright*, Cambridge (MA): MIT Press, 175–214.

Cartwright R., 1975, 'Scattered Objects', in K. Lehrer (ed.), *Analysis and Metaphysics*, Dordrecht: Reidel, 153–171.

Casati R., 1994, 'The Nature of Parts', Centre de Recherche in Epistemologie Appliquée (CREA), Paris, Lecture series on Language and Mind, February 1994.

Casati R., 1995a, 'Temporal Entities in Space', in P. Amsili, M. Borillo, and L. Vieu (eds.), *Time, Space and Movement: Meaning and Knowledge in the Sensible World. Proceedings of the 5th International Workshop*, Toulouse: COREP, Part D, 66–78.

Casati R., 1995b, 'The Shape Without', paper presented at the workshop *Topology and Dynamics in Cognition and Perception*, International Center for Semiotic and Cognitive Studies, San Marino, December 1995.

Casati R., 1999a, 'Formal Structures in the Phenomenology of Movement', in B. Pachoud, J. Petitot, J.-M. Roy, and F. Varela (eds.), *Naturalizing Phenomenology: Issues in Contemporary Phenomenology and Cognitive Science*, Stanford: Stanford University Press, forthcoming.

Casati R., 1999b, 'Formal Maps', *Journal of Semantics*, forthcoming.

Casati R. and Soldati, G., 1995, 'On the Perception of Abstract Objects', in J. Hill and P. Koťátko (eds.), *Karlovy Vary Studies in Reference and Meaning*, Prague: Filosofia, 89–113.

Casati R. and Varzi A. C., 1994, *Holes and Other Superficialities*, Cambridge (MA): MIT Press (Bradford Books).

Casati R. and Varzi A. C., 1995, 'Basic Issues in Spatial Representation', in M. De Glas and Z. Pawlak (eds.), *Proceedings of the 2nd World Conference on the Fundamentals of Artificial Intelligence*, Paris: Angkor, 63–72.

Casati R. and Varzi A. C., 1996, 'The Structure of Spatial Localization', *Philosophical Studies* 82, 205–239.

Casati R. and Varzi A. C., 1997a, *Fifty Years of Events: Annotated Bibliography 1947 to 1997*, Bowling Green (OH): Philosophy Documentation Center.

Casati R. and Varzi A. C., 1997b, 'Spatial Entities', in O. Stock (ed.), *Spatial and Temporal Reasoning*, Dordrecht: Kluwer, 73–96.

Casati R. and Varzi A. C., 1999, 'Topological Essentialism', *Philosophical Studies*, forthcoming.

Casati R. and Varzi A. C. (eds.), 1996, *Events*, Aldershot: Dartmouth.

Cave C. B. and Kosslyn S. M., 1993, 'The Role of Parts and Spatial Relations in Object Identification', *Perception* 22, 229–248.

Chisholm R. M., 1973, 'Parts as Essential to Their Wholes', *Review of Metaphysics* 26, 581–603.

Chisholm R. M., 1975, 'Mereological Essentialism: Some Further Considerations', *Review of Metaphysics* 28, 477–484.

Chisholm R. M., 1976, *Person and Object. A Metaphysical Study*, La Salle (IL): Open Court.

Chisholm R. M., 1984, 'Boundaries as Dependent Particulars', *Grazer philosophische Studien* 10, 87–95.

Chisholm R. M., 1987, 'Scattered Objects', in J. J. Thomson (ed.), *On Being and Saying: Essays for Richard Cartwright*, Cambridge (MA): MIT Press, 167–173.

Chisholm R. M., 1992/1993, 'Spatial Continuity and the Theory of Part and Whole. A Brentano Study', *Brentano Studien* 4, 11–23.

Chisholm R. M., 1994, 'Ontologically Dependent Entities', *Philosophy and Phenomenological Research* 54, 499–507.

Chomsky N., 1965, *Aspects of the Theory of Syntax*, Cambridge (MA): MIT Press.

Clarke B. L., 1981, 'A Calculus of Individuals Based on "Connection"', *Notre Dame Journal of Formal Logic* 22, 204–218.

Clarke B. L., 1985, 'Individuals and Points', *Notre Dame Journal of Formal Logic* 26, 61–75.

Clay R. E., 1974, 'Relation of Leśniewski's Mereology to Boolean Algebras', *Journal of Symbolic Logic* 39, 638–648.

Clay R. E., 1981, *Leśniewski's Mereology*, Cumana: Universidad de Oriente.

Cohn A. G., 1995, 'Qualitative Shape Representation using Connection and Convex Hulls', in P. Amsili, M. Borillo, and L. Vieu (eds.), *Time, Space and Movement: Meaning and Knowledge in the Sensible World. Proceedings of the 5th International Workshop*, Toulouse: COREP, Part C, 3–16.

Cohn A. G., Bennett B., Gooday J., and Gotts N. M., 1997, 'Representing and Reasoning with Qualitative Spatial Relations', in O. Stock (ed.), *Spatial and Temporal Reasoning*, Dordrecht: Kluwer, 97–134.

Cohn A. G. and Gotts N. M., 1996, 'The "Egg-Yolk" Representation of Regions with Indeterminate Boundaries', in P. A. Burrough and A. U. Frank (eds.), *Geographic Objects with Indeterminate Boundaries*, London: Taylor and Francis, 171–187.

Cohn A. G., Randell D. A., and Cui Z., 1995, 'A Taxonomy of Logically Defined Qualitative Spatial Regions', *International Journal of Human-Computer Studies* 43, 831–846.

Cohn A. G. and Varzi A. C., 1998, 'Connection Relations in Mereotopology', in H. Prade (ed.), *Proceedings of the 13th European Conference on Artificial Intelligence*, Chichester: Wiley, 150–154.

Copeland B. J., 1995, 'On Vague Objects, Fuzzy Logic and Fractal Boundaries', *Southern Journal of Philosophy* 33 (Supplement), 83–96.

Couclelis H., 1996, 'Typology of Geographic Entities with Ill-Defined Boundaries', in P. A. Burrough and A. U. Frank (eds.), *Geographic Objects with Indeterminate Boundaries*, London: Taylor and Francis, 45–56.

Cruse D. A., 1979, 'On the Transitivity of the Part-Whole Relation', *Journal of Linguistics* 15, 29–38.

Davidson D., 1969, 'The Individuation of Events', in N. Rescher (ed.), *Essays in Honor of Carl G. Hempel*, Dordrecht: Reidel, 216–234.

Davidson D., 1985, 'Reply to Quine on Events', in E. LePore and B. McLaughlin (eds.), *Actions and Events: Perspectives on the Philosophy of Donald Davidson*, Oxfod: Blackwell, 172–176.

Davis E., 1987, 'A Framework for Qualitative Reasoning About Solid Objects', in G. Rodriguez (ed.), *Proceedings of the Workshop on Space Telerobotics*, Pasadena (CA): NASA and JPL, 369–375.

Davis E., 1993, 'The Kinematics of Cutting Solid Objects', *Annals of Mathematics and Artificial Intelligence* 9, 253–305.

De Laguna T., 1922a, 'The Nature of Space—I', *Journal of Philosophy* 19, 393–407.

De Laguna T., 1922b, 'The Nature of Space—II. The Empirical Basis of Geometry', *Journal of Philosophy* 19, 421–440.

De Laguna T., 1922c, 'Point, Line, and Surface, as Sets of Solids', *Journal of Philosophy* 19, 449–461.

Della Rocca M., 1996, 'Essentialists and Essentialism', *Journal of Philosophy* 93, 186–202.

Desclés J.-P., 1989, 'State, Event, Process, and Topology', *General Linguistics* 29, 159–200.

Doepke F. C., 1982, 'Spatially Coinciding Objects', *Ratio* 24, 45–60.

Doepke F. C., 1986, 'In Defence of Locke's Principle: A Reply to Peter M. Simons', *Mind* 95, 238–241.

Dretske F., 1961, 'Particulars and the Relational Theory of Time', *The Philosophical Review* 70, 447–469.

Dretske F., 1962, 'Moving Backward in Time', *The Philosophical Review* 71, 94–98.

Dretske F., 1967, 'Can Events Move?', *Mind* 76, 479–492.

Earman J., 1989, *World Enough and Space-Time. Absolute versus Relational Theories of Space and Time*, Cambridge (MA): MIT Press.

Eberle R. A., 1967, 'Some Complete Calculi of Individuals', *Notre Dame Journal of Formal Logic* 8, 267–278.

Eberle R. A., 1970, *Nominalistic Systems*, Dordrecht: Reidel.

Egenhofer M. J., 1991, 'Reasoning about Binary Topological Relations', in O. Günther and H.-J. Schek (eds.) *Advances in Spatial Databases*, Berlin: Springer, 143–160.

Egenhofer M. J. and Franzosa R. D., 1991, 'Point-Set Topological Spatial Relations', *International Journal of Geographical Information Systems* 5, 161–174.

Egenhofer M. J. and Franzosa R. D., 1995, 'On the Equivalence of Topological Relations', *International Journal of Geographical Information Systems* 9, 133–152.

Egenhofer M. J. and Mark D., 1995, 'Naive Geography', in A. U. Frank and W. Kuhn (eds.), *Spatial Information Theory: A Theoretical Basis for GIS. Proceedings of the Third International Conference*, Berlin: Springer, 1–15.

Eilan N., McCarthy R., and Brewer B. (eds.), 1993, *Spatial Representation. Problems in Philosophy and Psychology*, Oxford: Blackwell.

Eschenbach C., 1994, 'A Mereotopological Definition of "Point"', in C. Eschenbach, C. Habel and B. Smith (eds.), *Topological Foundations of Cognitive Science. Papers from the Workshop at the First International Summer Institute in Cognitive Science*, University of Hamburg, Reports of the Doctoral Program in Cognitive Science, No. 37, 63–80.

Eschenbach C. and Heydrich W., 1995, 'Classical Mereology and Restricted Domains', *International Journal of Human-Computer Studies* 43, 723–740.

Evans G., 1978, 'Can There Be Vague Objects?', *Analysis* 38, 208.

Fine K., 1981, 'Model Theory for Modal Logic, Part III: Existence and Predication', *Journal of Philosophical Logic* 10, 293–237.

Fine K., 1994, 'A Puzzle Concerning Matter and Form', in T. Scaltsas, D. Charles, and M. L. Gill (eds.), *Unity, Identity and Explanation in Aristotle's Metaphysics*, Oxford: Clarendon, 13–40.

Fine K., 1994, 'Compounds and Aggregates', *Noûs* 28, 137–158.

Fine K., 1995a, 'Part–Whole', in B. Smith and D. W. Smith (eds.), *The Cambridge Companion to Husserl*, Cambridge: Cambridge University Press, 463–485.

Fine K., 1995b, 'Ontological Dependence', *Proceedings of the Aristotelian Society* 95, 269–290.

Forrest P., 1984, 'Is Motion Change of Location?', *Analysis* 44, 177–178.

Forrest P., 1996a, 'From Ontology to Topology in the Theory of Regions', *The Monist* 79, 34–50.

Forrest P., 1996b, 'How Innocent Is Mereology?', *Analysis* 56, 127–131.

Frank A. U., 1997, 'Spatial Ontology: A Geographical Information View', in O. Stock (ed.), *Spatial and Temporal Reasoning*, Dordrecht: Kluwer, 135–153.

Frege G., 1884, *Die Grundlagen der Arithmetik. Eine logisch-mathematische Untersuchung über den Begriff der Zahl*, Breslau: Köbner (Eng. trans. by J. L. Austin, *Foundations of Arithmetic*, Oxford: Blackwell, 1950).

Galton A. P., 1993, 'Towards an Integrated Logic of Space, Time, and Motion', in *Proceedings of the 13th International Joint Conference on Artificial Intelligence*, Chambéry: IJCAI [Morgan Kaufmann], Vol. 2, 1550–1555.

Galton A. P., 1994, 'Instantaneous Events', in H. J. Ohlbach (ed.), *Temporal Logic: Proceedings of the ICTL Workshop*, Saarbrücken: Max-Planck-Institut für Informatik, Technical Report MPI-I-94-230, 4–11.

Galton A. P., 1996, 'Time and Continuity in Philosophy, Mathematics, and Artificial Intelligence', *Kodikas/Code* 19, 101–119.

Galton A., 1997, 'Space, Time, and Movement', in O. Stock (ed.), *Spatial and Temporal Reasoning*, Dordrecht: Kluwer, 321–352.

Galton A., 1998, 'Modes of Overlap', *Journal of Visual Languages and Computing* 9, 61–79.

Galton A. P., 1999, *Qualitative Spatial Change*, forthcoming.

Garrett B. J., 1985, 'Noonan, "Best Candidate" Theories, and the Ship of Theseus', *Analysis* 45, 12–15.

Geach P. T., 1967, 'Identity', *Review of Metaphysics* 21, 3–12.

Geach P. T., 1980, *Reference and Generality*, Third Edition, Ithaca: Cornell University Press.

Geach P. T., 1982a, 'Reply to Lowe', *Analysis* 42, 30.

Geach P. T., 1982b, 'Reply to Lowe's Reply', *Analysis* 42, 32.

Gerla G. and Tortora R., 1992, 'La relazione di connessione in A. N. Whitehead: aspetti matematici', *Epistemologia* 15, 351–364.

Gerla G., 1995, 'Pointless Geometries', Chapter 18 of F. Buekenhout (ed.), *Handbook of Incidence Geometry*, Amsterdam: Elsevier, 1015–1031.

Gerstl P. and Pribbenow S., 1995, 'Midwinters, End Games, and Bodyparts. A Classification of Part-Whole Relations', *International Journal of Human-Computer Studies* 43, 865–889.

Gibson J. J., 1966, *The Senses Considered as Perceptual Systems*, London: Allen & Unwin.

Gibson J. J., 1979, *The Ecological Approach to Visual Perception*, Boston: Houghton Mifflin.

Giorgi A. and Pianesi F., 1997, *Tense and Aspect. From Semantics to Morphosyntax*, Oxford: Oxford University Press.

Godfrey R., 1990, 'Democritus and the Impossibility of Collision', *Philosophy* 65, 212–217.

Goodman N., 1951, *The Structure of Appearance*, Cambridge (MA): Harvard University Press (3rd ed. Dordrecht: Reidel, 1977).

Goodman N., 1956, 'A World of Individuals', in J. M. Bocheński, A. Church, and N. Goodman, *The Problem of Universals. A Symposium*, Notre Dame: University of Notre Dame Press, 13–31.

Goodman N., 1958, 'On Relations that Generate', *Philosophical Studies* 9, 65–66.

Gotts N. M., 1994a, 'How Far Can We '**C**'? Defining a 'Doughnut' Using Connection Alone', in J. Doyle, E. Sandewall, and P. Torasso (eds.), *Principles of Knowledge Representation and*

Reasoning: Proceedings of the Fourth International Conference, San Mateo (CA): Morgan Kaufmann, 246–257.

Gotts N. M., 1994b, 'Defining a 'Doughnut' Made Difficult', in C. Eschenbach, C. Habel and B. Smith (eds.), *Topological Foundations of Cognitive Science. Papers from the Workshop at the First International Summer Institute in Cognitive Science*, University of Hamburg, Reports of the Doctoral Program in Cognitive Science, No. 37, 105–129.

Gotts N. M., Gooday J. M., and Cohn A. G., 1996, 'A Connection Based Approach to Common-Sense Topological Description and Reasoning', *The Monist* 79, 51–75.

Grzegorczyk A., 1960, 'Axiomatizability of Geometry Without Points', *Synthese* 12, 109–127.

Griffin N., 1977, *Relative Identity*, Oxford: Clarendon.

Habel C., 1989, 'Zwischen-Bericht', in C. Habel, M. Herweg, and K. Rehkämper (eds.), *Raumkonzepte in Verstehenprozessen*, Tübingen: Niemeyer, 37–69.

Hacker P. M. S., 1982, 'Events and Objects in Space and Time', *Mind* 91, 1–19.

Hamblin C., 1969, 'Starting and Stopping', *The Monist* 53, 410–425.

Hamblin C., 1971, 'Instants and Intervals', *Studium Generale* 24, 127–134.

Haslanger S., 1994, 'Parts, Compounds, and Substantial Unity', in T. Scaltsas, D. Charles, and M. L. Gill (eds.), *Unity, Identity and Explanation in Aristotle's Metaphysics*, Oxford: Clarendon, 129–170.

Hayes P. J., 1979, 'The Naive Physics Manifesto', in D. Michie (ed.), *Expert Systems in the Micro-Electronic Age*, Edinburgh: Edinburgh University Press, 242–270.

Hayes P. J., 1985a, 'The Second Naive Physics Manifesto', in R. Hobbs and R. C. Moore (eds.), *Formal Theories of the Commonsense World*, Norwood: Ablex, 1–36.

Hayes P. J., 1985b, 'Naive Physics I: Ontology for Liquids', in J. R. Hobbs and R. C. Moore (eds.), *Formal Theories of the Commonsense World*, Norwood: Ablex, 71–107.

Hazen A. P., 1990, 'The Mathematical Philosophy of Contact', *Philosophy* 65, 205–211.

Hazen A. P., 1993, 'Slicing It Thin', *Analysis* 53, 189–192.

Heider F., 1926, 'Ding und Medium', *Symposion* 1, 109–157 (republished as 'Thing and Medium', *Psychological Issues* 1 (1959), 1–34.)

Heller M., 1984, 'Temporal Parts of Four Dimensional Objects', *Philosophical Studies* 46, 323–334.

Heller M., 1990, *The Ontology of Physical Objects: Four-Dimensional Hunks of Matter*, Cambridge: Cambridge University Press.

Heller M., 1996, 'Against Metaphysical Vagueness', *Philosophical Perspectives* 10, 177–186.

Hempel C. G., 1953, 'Reflections on Nelson Goodman's "The Structure of Appearance"', *The Philosophical Review* 62, 108–116.

Henry D., 1991, *Medieval Mereology*, Amsterdam: Grüner.

Henry D., 1994, 'Impenetrability, Overlapping, and Connumeration', *Axiomathes* 5:1, 77–86.

Herskovits A., 1986, *Language and Spatial Cognition. An Interdisciplinary Study of the Prepositions in English*, Cambridge: Cambridge University Press.

Herskovits A., 1997, 'Language, Spatial Cognition, and Vision', in O. Stock (ed.), *Spatial and Temporal Reasoning*, Dordrecht: Kluwer, 155–202.

Herweg M., 1989, 'Ansätze zu einer semantischen Beschreibung topologischer Präpositionen', in C. Habel, M. Herweg, and K. Rehkämper (eds.), *Raumkonzepte in Verstehenprozessen*, Tübingen: Niemeyer, 99–127.

Hestevold H. S., 1986, 'Boundaries, Surfaces, and Continuous Wholes', *Southern Journal of Philosophy* 24, 235–245.

Heydrich W., 1988, 'Things in Space and Time', in J. S. Petöfi (ed.), *Text and Discourse Constitution. Empirical Aspects, Theoretical Approaches*, Berlin: de Gruyter, 377–418.

Hexmoor H. H., 1990, 'A Calculus for Assembly', in G. Rzevski (ed.), *Applications of Artificial Intelligence in Engineering V*, Berlin: Springer, 427–446.

Hirsh E., 1982, *The Concept of Identity*, Oxford: Oxford University Press.

Hoffman D. and Richards W. A., 1985, 'Parts of Recognition', *Cognition* 18, 65–96.

Hoffman J. and Rosenkrantz G. S., 1997, *Substance. Its Nature and Existence*, London: Routledge.

Horwich P., 1977, 'On the Existence of Times, Spaces, and Space-Times', *Noûs* 12, 396–419.

Husserl E., 1900/1901, *Logische Untersuchungen. Zweiter Band. Untersuchungen zur Phänomenologie und Theorie der Erkenntnis*, Halle: Niemeyer (2nd ed. 1913; Eng. trans. by J. N. Findlay, *Logical Investigations, Volume Two*, London: Routledge & Kegan Paul, 1970).

Iris M. A., Litowitz B. E., and Evens M., 1988, 'Problems of the Part-Whole Relation', in M. Evens (ed.), *Relations Models of the Lexicon*, Cambridge: Cambridge University Press, 261–288.

Jackendoff R., 1991, 'Parts and Boundaries', *Cognition* 41, 9–45.

Jackson F., 1977, *Perception. A Representative Theory*, Cambridge: Cambridge University Press.

Jansana R., 1994, 'Some Logics Related to von Wright's Logic of Place', *Notre Dame Journal of Formal Logic* 35, 88–98.

Johansson I., 1989, *Ontological Investigations. An Inquiry into the Categories of Nature, Man and Society*, London: Routledge.

Kamp H., 1979, 'Events, Instants, and Temporal Reference', in R. Bäuerle, U. Egli and A. von Stechow (eds.), *Semantics from Different Points of View*, Berlin: Springer, 376–417.

Kim J., 1976, 'Events as Property Exemplifications', in M. Brand and D. Walton (eds.), *Action Theory*, Dordrecht: Reidel, 159–177.

Kline A. D. and Matheson C. A., 1987, 'The Logical Impossibility of Collision', *Philosophy* 62, 509–515.

Kosslyn S. M., Heldmeyer K. H., and Glass A. L., 1980, 'Where Does One Part End and Another Begin? A Developmental Study', in J. Becker, F. Wilkening, and T. Trabasso (eds.), *Information Integration in Children*, Hillsdale (NJ): Erlbaum, 147–168.

Krifka M., 1989, 'Four Thousand Ships Passed Through the Lock: Object-Induced Measure Functions on Events', *Linguistics and Philosophy* 13, 487–520.

Kripke S., 1972, 'Naming and Necessity', in D. Davidson and G. Harman (eds.), *Semantics of Natural Language*, Dordrecht: Reidel, 253–355, addenda 763–769 (reprinted as *Naming and Necessity*, Cambridge (MA): Harvard University Press, 1980).

Künne W., 1983, *Abstrakte Gegenstände*, Frankfurt: Suhrkamp.

Kuratowski C., 1922, 'Sur l'opération A- de l'Analysis Situs', *Fundamenta Mathematicae* 3, 182–199.

Lacey H. and Anderson E., 1980, 'Spatial Ontology and Spatial Modalities', *Philosophical Studies* 38, 261–285.

Lacey H., 1971, 'The Philosophical Intellegibility of Absolute Space: A Study of Newtonian Argument', *British Journal for the Philosohpy of Science* 21, 317–342.

Landau B. and Jackendoff R., 1993, '"What" and "Where" in Spatial Language and Spatial Cognition', *Behavioral and Brain Sciences* 16, 217–265.

Landman F., 1991, *Structures for Semantics*, Dordrecht: Kluwer.

Lejewski C., 1982, 'Ontology: What's Next?', in W. Leinfellner, E. Kraemer, and J. Schank (eds.), *Language and Ontology. Proceedings of the 6th International Wittgenstein Symposium*, Vienna: Hölder-Pichler-Tempsky, 173–185.

Lejewski C., 1986, 'Logic, Ontology and Metaphysics', in S. G. Shanker (ed.), *Philosophy in Britain Today*, London: Croom Helm, 171–197.

Lemmon E. J., 1967, 'Comments on D. Davidson's "The Logical Form of Action Sentences"', in N. Rescher (ed.), *The Logic of Decision and Action*, Pittsburgh: Pittsburgh University Press, 96–103.

Lemon O., 1996, 'Semantical Foundations of Spatial Logics', in L. C. Aiello, J. Doyle, and S. C. Shapiro (eds.), *Principles of Knowledge Representation and Reasoning. Proceedings of the Fifth International Conference*, San Mateo (CA): Morgan Kaufmann, 212–219.

Lemon O. and Pratt I., 1998, 'Complete Logics for QSR: A Guide to Plane Mereotopology', *Journal of Visual Languages and Computing* 9, 5–21.

Leonard H. S. and Goodman N., 1940, 'The Calculus of Individuals and Its Uses', *Journal of Symbolic Logic* 5, 45–55.

Leong M. K., 1994, *Towards a Semantics for a Visual Information System*, PhD Dissertation, Stanford University.

Leśniewski S., 1916, *Podstawy ogólnej teoryi mnogości*. I, Moskow: Prace Polskiego Koła Naukowego w Moskwie, Sekcya matematyczno-przyrodnicza (Eng. trans. by D. I. Barnett, 'Foundations of the General Theory of Sets. I', in S. Leśniewski, *Collected Works*, ed. S. J. Surma, J. Srzednicki, D. I. Barnett, and F. V. Rickey, Dordrecht: Kluwer, 1992, Vol. 1, 129–173).

Leśniewski S., 1927/1931, 'O podstawach matematyki', *Przegląd Filozoficzny* 30, 164–206; 31, 261–291; 32, 60–101; 33, 77–105; 34, 142–170 (Eng. trans. by D. I. Barnett, 'On the Foundations of Mathematics', in S. Leśniewski, *Collected Works*, ed. S. J. Surma, J. T. Srzednicki, D. I. Barnett, and F. V. Rickey, Dordrecht: Kluwer, 1992, Vol. 1, 174–382).

Levey S., 1997, 'Coincidence and Principles of Composition', *Analysis* 57, 1–10.

Lewis D. K., 1968, 'Counterpart Theory and Quantified Modal Logic', *Journal of Philosophy* 65, 113–126.

Lewis D. K., 1970, 'Nominalistic Set Theory', *Noûs* 4, 225–240.

Lewis D. K., 1986, *The Plurality of Worlds*, Oxford: Blackwell.

Lewis D. K., 1991, *Parts of Classes*, Oxford: Blackwell.

Lewis D. K., 1993a, 'Mathematics is Megethology', *Philosophia Mathematica* 3, 3–23.

Lewis D. K., 1993b, 'Many, but Almost One', in J. Bacon, K. Campbell, and L. Reinhardt (eds.), *Ontology, Causality, and Mind: Essays in Honor of D. M. Armstrong*, Cambridge: Cambridge University Press, 23–38.

Lewis D. K. and Lewis S. R., 1970, 'Holes', *Australasian Journal of Philosophy* 48, 206–212.

Libardi M., 1994, 'Applications and Limits of Mereology. From the Theory of Parts to the Theory of Wholes', *Axiomathes* 5:1, 13–54.

Link G., 1987, 'Algebraic Semantics for Event Structures', in J. Groenendijk, M. Stockhof, and F. Veltman (eds.), *Proceedings of the 6th Amsterdam Colloquium*, Amsterdam: Institute for Language, Logic and Information, 243–262.

Litowitz B. and Novy F., 1984, 'Expression of the Part-Whole Semantic Relation by 3 to 12 Year Old Children', *Journal of Child Language* 11, 159–178.

Lombard L. B., 1986, *Events. A Metaphysical Study*, London: Routledge & Kegan Paul.

Lowe E. J., 1982a, 'The Paradox of the 1,001 Cats', *Analysis* 42, 27–30.

Lowe E. J., 1982b, 'Reply to Geach', *Analysis* 42, 31.

Lowe E. J., 1982c, 'On Being a Cat', *Analysis* 42, 174–177.

Lowe E. J., 1989, *Kinds of Being*, Oxford: Blackwell.

Lowe E. J., 1995, 'Coinciding Objects: In Defence of the "Standard Account"', *Analysis* 55, 171–178.

Lowe V., 1953, 'Professor Goodman's Concept of an Individual', *Philosophical Review* 62, 117–126.

Luschei E. C., 1965, *The Logical Systems of Leśniewski*, Amsterdam: North-Holland.

Lyons J., 1977, *Semantics, Volume I*, Cambridge: Cambridge University Press.

Marr D. and Nishihara H. K., 1977, 'Representation and Recognition of the Spatial Organization of Three-Dimensional Shapes', *Proceedings of the Royal Society*, B 200, 269–294.

Martin C. B., 1996, 'How It Is: Entities, Absences and Voids', *Australasian Journal of Philosophy* 74, 57–65.

Martin R. M., 1965, 'Of Time and the Null Individual', *Journal of Philosophy* 62, 723–736.

Martin R. M., 1978, *Events, Reference, and Logical Form*, Washington: Catholic University of America Press.

Mayo B., 1961, 'Objects, Events, and Complementarity', *Mind* 70, 340–361.

Mehlberg H., 1956, *The Reach of Science*, Toronto: Toronto University Press.

Mellor D. H., 1980, 'Things and Causes in Spacetime', *British Journal for the Philosophy of Science* 31, 282–288.

Mellor D. H., 1995, *The Facts of Causation*, London: Routledge.

Menger K., 1940, 'Topology Without Points', *Rice Institute Pamphlets* 27, 80–107.

Meyer R. K. and Lambert K., 1968, 'Universally Free Logic and Standard Quantification Theory', *Journal of Symbolic Logic* 33, 8–26.

Michaels M., 1983, 'Identity Through Time: One Last Run for the Ordinary View', *Pacific Philosophical Quarterly* 64, 97–109.

Miéville D., 1984, *Un développement des systèmes logiques de Stanislaw Leśniewski. Protothétique – Ontologie – Méréologie*, Berne: Lang.

Moltmann F., 1991, 'The Multidimensional Part Structure of Events', in A. L. Halpern (ed.), *Proceedings of the Ninth West Coast Conference on Fornal Linguistics*, Stanford: Center for the Study of Language and Information, 361–378.

Moltmann F., 1997, *Parts and Wholes in Semantics*, Oxford: Oxford University Press.

Monmonier M., 1991, *How to Lie With Maps*, Chicago: University of Chicago Press.

Mortensen C. and Nerlich G., 1978, 'Physical Topology', *Journal of Philosophical Logic* 7, 209–223.

Mulligan K, 1993, 'Internal Relations', in B. Garrett and P. Menzies (eds.), *Proceedings of the 1992 ANU Metaphysics Conference*, Australian National University, 1–22.

Neale S., 1990, *Descriptions*, Cambridge (MA): MIT Press (Bradford Books).

Needham P., 1981, 'Temporal Intervals and Temporal Order', *Logique et Analyse* 24, 49–64.

Nerlich G., 1994, *The Shape of Space*, Second Edition, Cambridge: Cambridge University Press.

Newton-Smith W. H., 1980, *The Structure of Time*, London: Routledge & Kegan Paul.

Nicod J., 1924, *La géométrie dans le monde sensible*, Paris: Alcan (Eng. trans. by J. Bell, 'Geometry in the Sensible World', in J. Nicod, *Geometry and Induction*, ed. R. F. Harrod, London: Routledge & Kegan Paul, 1970, 3–155).

Noonan H. W., 1980, *Objects and Identity*, The Hague: Nijhoff.

Noonan H. W., 1986, 'Reply to Simons on Coincidence', *Mind* 95, 238–241.

Noonan H. W., 1993, 'Constitution Is Identity', *Mind* 102, 133–146.

Noonan H. W. (ed.), 1993, *Identity*, Aldershot: Dartmouth.

Null G., 1983, 'A First-Order Axiom System for Non-Universal Part-Whole and Foundation Relations', in L. Embree (ed.), *Essays in Memory of Aron Gurwitsch*, Lanham (MD): University Press of America, 463–484.

Oderberg D. S., 1996, 'Coincidence Under a Sortal', *The Philosophical Review* 105, 145–171.

Olson E. T., 1996, 'Composition and Coincidence', *Pacific Philosophical Quarterly* 77, 374–403.

Peirce C. S., 1893, 'The Logic of Quantity', in *Collected Papers of Charles Sanders Peirce*, Vol. IV, ed. C. Hartshorne and P. Weiss, Cambridge (MA): Harvard University Press, 1933.

Perzanowski J., 1993, 'Locative Ontology. Parts I–III', *Logic and Logical Philosophy* 1, 7–94.

Pianesi F. and Varzi A. C., 1994a, 'The Mereo-Topology of Event Structures', in P. Dekker and M. Stokhof (eds.), *Proceedings of the 9th Amsterdam Colloquium*, Amsterdam: Institute for Logic, Language and Computation, 527–546.

Pianesi F. and Varzi A. C., 1994b, 'Mereotopological Construction of Time from Events', in A. G. Cohn (ed.), *Proceedings of the 11th European Conference on Artificial Intelligence*, Chichester: Wiley, 396–400.

Pianesi F. and Varzi A. C., 1996a, 'Events, Topology, and Temporal Relations', *The Monist* 78, 89–116.

Pianesi F. and Varzi A. C., 1996b, 'Refining Temporal Reference in Event Structures', *Notre Dame Journal of Formal Logic* 37, 71–83.

Piñón C., 1997, 'Achievements in an Event Semantics', in A. Lawson and E. Cho (eds.), *Proceedings from the Seventh Conference on Semantics and Linguistic Theory*, Ithaca (NY): CLC Publications, Cornell University.

Plantinga A., 1975, 'On Mereological Essentialism', *Review of Metaphysics* 27, 468–484.

Polkowski L. and Skowron A., 1994, 'Rough Mereology', in Z. W. Ras and M. Zemankova (eds.), *Proceedings of the 8th International Symposium on Methodologies for Intelligent Systems*, Berlin: Springer, 85–94.

Pratt I., 1993, 'Map Semantics', in A Frank and I. Campari (eds.), *Spatial Information Theory: A Theoretical Basis for GIS. Proceedings of the Second International Conference*, Berlin: Springer, 77–91.

Pratt I. and Lemon O., 1997, 'Ontologies for Plane, Polygonal Mereotopology', *Notre Dame Journal of Formal Logic* 38, 225–245.

Price H. H., 1932, *Perception*, London: Methuen.

Putnam H., 1987, *The Many Faces of Realism*, La Salle (IL): Open Court.

Quine W. V. O., 1950, 'Identity, Ostension and Hyposthasis', *Journal of Philosophy* 47, 621–633.

Quine W. V. O., 1960, *Word and Object*, Cambridge (MA): MIT Press.

Quine W. V. O., 1985, 'Events and Reification', in E. LePore and B. P. McLaughlin (eds.), *Actions and Events. Perspectives in the Philosophy of Donald Davidson*, Oxford: Blackwell, 162–171.

Quine W. V. O., 1970, *Philosophy of Logic*, Englewood Cliffs (NJ): Prentice-Hall.

Quinton A., 1979, 'Objects and Events', *Mind* 88, 197–214.

Randell D. A., 1991, *Analysing the Familiar: Reasoning about Space and Time in the Everyday World*, University of Warwick: PhD Thesis.

Randell D. A. and Cohn A. G., 1989, 'Modelling Topological and Metrical Properties in Physical Processes', in R. J. Brachman, H. J. Levesque and R. Reiter (eds.), *Principles of Knowledge Representation and Reasoning. Proceedings of the First International Conference*, Los Altos (CA): Morgan Kaufmann, 357–368.

Randell D. A. and Cohn A. G., 1992, 'Exploiting Lattices in a Theory of Space and Time', *Computers and Mathematics with Applications* 23, 459–476.

Randell D. A., Cui Z., and Cohn A. G., 1992a, 'An Interval Logic of Space Based on "Connection"', in B. Neumann (ed.), *Proceedings of the 10th European Conference on Artificial Intelligence*, Chichester: Wiley, 394–398.

Randell D. A., Cui Z., and Cohn A. G., 1992b, 'A Spatial Logic Based on Regions and Connections', in B. Nebel, C. Rich, and W. Swartout (eds.), *Principles of Knowledge Representation and Reasoning. Proceedings of the Third International Conference*, Los Altos (CA): Morgan Kaufmann, 165–176.

Ray C., 1991, *Time, Space and Philosophy*, London: Routledge.

Rea M., 1995, 'The Problem of Material Constitution', *The Philosophical Review* 104, 525–547.

Rea M., 1997, 'Supervenience and Co-Location', *American Philosophical Quarterly* 34, 367–375.

Rea M., 1998, 'In Defense of Mereological Universalism', *Philosophy and Phenomenological Research* 58, 347–360.

Rea M. (ed.), 1997, *Material Constitution. A Reader*, Lanham (MD): Rowman & Littlefield.

Rescher N., 1955, 'Axioms for the Part Relation', *Philosophical Studies* 6, 8–11.

Reuleaux F., 1875, *Theoretische Kinematik: Grundzüge einer Theorie des Maschinenwesens*, Braunschweig: Vieweg (Eng. trans. by A. B. W. Kennedy, *The Kinematics of Machinery: Outline of a Theory of Machines*, London: Macmillan, 1876; reprinted by Dover, 1963).

Robinson A. H. and Petchenik B. B., 1976, *The Nature of Maps: Essays Towards Understanding Maps and Mapping*, Chicago: University of Chicago Press.

Robinson D., 1982, 'Re-Identifying Matter', *The Philosophical Review* 91, 317–341.

Roeper P., 1997, 'Region-Based Topology', *Journal of Philosophical Logic* 26, 251–309.

Russell B. A. W., 1905, 'On Denoting', *Mind* 14, 479–493.

Russell B. A. W., 1914, *Our Knowledge of the External World*, London: Allen & Unwin.

Russell B. A. W., 1927, *The Analysis of Matter*, London: Allen & Unwin.

Russell B. A. W., 1940, *An Inquiry into Meaning and Truth*, London: Allen & Unwin.

Sainsbury R. M., 1995, 'Why the World Cannot Be Vague', *Southern Journal of Philosophy* 33 (Supplement), 63–81.

Salmon N., 1997, 'Wholes, Parts, and Numbers', *Philosophical Perspectives* 11, 1–15.

Sanford D., 1993, 'The Problem of the Many, Many Composition Questions, and Naive Mereology', *Noûs* 27, 219–228.

Sayan E., 1996, 'A Mereological Look at Motion', *Philosophical Studies* 84, 75–89.

Scaltsas T., 1990, 'Is a Whole Identical to Its Parts?', *Mind* 99, 583–598.

Schmolze J. G., 1996, 'A Topological Account of the Space Occupied by Physical Objects', *The Monist* 79, 128–140.

Sharvy R., 1983a, 'Aristotle on Mixtures', *Journal of Philosophy* 80, 439–457.

Sharvy R., 1983b, 'Mixtures', *Philosophy and Phenomenological Research* 44, 227–239.

Shoham Y., 1984, 'Naive Kinematics: Two Aspects of Shape', in J. R. Hobbs (ed.), *Commonsense Summer: Final Report*, Technical Report CSLI-85–35, Stanford: SRI International, AI Center, Part 4, 1–25.

Shorter J. M., 1977, 'On Coinciding in Space and Time', *Philosophy* 52, 399–408.

Sider T., 1997, 'Four-Dimensionalism', *The Philosophical Review* 106, 197–231.

Simons P. M., 1985, 'Coincidence of Things of a Kind', *Mind* 94, 70–75.

Simons P. M., 1986, 'Unkindly Coincidences', *Mind* 95, 506–509.

Simons P. M., 1987, *Parts. A Study in Ontology*, Oxford: Clarendon.

Simons P. M., 1991a, 'Part/Whole II: Mereology Since 1900', in H. Burkhardt and B. Smith (eds.), *Handbook of Metaphysics and Ontology*, Munich: Philosophia, 209–210.

Simons P. M., 1991b, 'Free Part-Whole Theory', in K. Lambert (ed.), *Philosophical Applications of Free Logic*, Oxford: Oxford University Press, 285–306.

Simons P. M., 1991c, 'Faces, Boundaries, and Thin Layers', in A. P. Martinich and M. J. White (eds.), *Certainty and Surface in Epistemology and Philosophical Method. Essays in Honor of Avrum Stroll*, Lewiston: Mellen, 87–99.

Simons P. M., 1991d, 'Whitehead und die Mereologie', in M. Hampe and H. Maassen (eds.), *Die Gifford Lectures und ihre Deutung. Materialien zu Whiteheads "Prozess und Realität"*, Vol. 2, Frankfurt: Suhrkamp, 369–388.

Simons P. M. and Dement C. W., 1996, 'Aspects of the Mereology of Artifacts', in R. Poli and P. M. Simons (eds.), *Formal Ontology*, Dordrecht: Kluwer, 255–276.

Smart J. J. C., 1972, 'Space-Time and Individuals', in R. Rudner and I. Schaeffer (eds.), *Logic and Art. Essays in Honor of Nelson Goodman*, New York: Macmillan, 3–20.

Smart J. J. C., 1982, 'Sellars on Process', *The Monist* 65, 302–314.

Smith B., 1982, 'Annotated Bibliography of Writings on Part-Whole Relations since Brentano', in B. Smith (ed.), *Parts and Moments. Studies in Logic and Formal Ontology*, Munich: Philosophia, 481–552.

Smith B., 1985, 'Addenda to: Annotated Bibliography of Writings on Part-Whole Relations since Brentano', in P. Sällström (ed.), *An Inventory of Present Thinking about Parts and Wholes*, vol. 3, Stockholm: Forskningsrådsnämnden, 74–86.

Smith B., 1992, '*Characteristica Universalis*', in K. Mulligan (ed.), *Language, Truth and Ontology*, Dordrecht: Kluwer, 50–81.

Smith B., 1993, 'Ontology and the Logistic Analysis of Reality', in N. Guarino and R. Poli (eds.), *International Workshop on Formal Ontology in Conceptual Analysis and Knowledge Representation*, Padova: CNR, 51–68 (revised and expanded version as 'Mereotopology: A Theory of Parts and Boundaries', *Data & Knowledge Engineering* 20 (1996), 287–304).

Smith B., 1994, 'Fiat Objects', in N. Guarino, S. Pribbenow, and L. Vieu (eds.), *Parts and Wholes: Conceptual Part-Whole Relations and Formal Mereology. Proceedings of the ECAI94 Workshop*, Amsterdam: European Coordinating Commettee for Artificial Intelligence, 15–23.

Smith B., 1995a, 'Zur Kognition räumlicher Grenzen: Eine mereotopologische Untersuchung', *Kognitionswissenschaft* 4:4, 177–184.

Smith B., 1995b, 'On Drawing Lines on a Map', in A. U. Frank and W. Kuhn (eds.), *Spatial Information Theory. A Theoretical Basis for GIS. Proceedings of the Third International Conference*, Berlin: Springer, 475–484.

Smith B., 1995c, 'More Things in Heaven and Earth', *Grazer Philosophisce Studien* 50, 187–201.

Smith B., 1997a, 'On Substances, Accidents, and Universals: In Defence of a Constituent Ontology', *Philosophical Papers* 26, 105–127.

Smith B., 1997b, 'Boundaries: An Essay in Mereotopology', in L. H. Hahn (ed.), *The Philosophy of Roderick Chisholm*, La Salle (IL): Open Court, 534–561.

Smith B., 1998, 'Basic Concepts of Formal Ontology', in N. Guarino (ed.), *Formal Ontology in Information Systems*, Amsterdam: IOS Press, 19–28.

Smith B. and Varzi A. C., 1999a, 'Fiat and Bona Fide Boundaries', *Philosophy and Phenomenological Research*, forthcoming.

Smith B. and Varzi A. C., 1999b, 'The Niche', *Noûs*, forthcoming.

Smullyan A. F., 1948, 'Modality and Description', *Journal of Symbolic Logic* 13, 31–37.

Sobociński B., 1954, 'Studies in Leśniewski's Mereology', *Polskie Towarzystwo Naukowe Na Obczyźnie*, 5, 34–48.

Sorabji R., 1983, *Time, Creation, and the Continuum: Theories in Antiquity and the Early Middle Ages*, Ithaca (NY): Cornell University Press.

Sorabji R., 1988, *Matter, Space, and Motion: Theories in Antiquity and Their Sequel*, Ithaca (NY): Cornell University Press.

Sorensen R. A., 1986, 'Transitions', *Philosophical Studies* 50, 187–193.

Sorensen R. A., 1998, 'Sharp Boundaries for Blobs', *Philosophical Studies* 91, 275–295.

Spelke E. S., Berlinger K., Jacobson K., and Phillips A., 1993, 'Gestalt Relations and Object Perception: A Developmental Study', *Perception* 22, 1483–1501.

Stell J. G., 1997, 'A Lattice-Theoretic Account of Spatial Relations', Technical Report, Keele University, Department of Computer Science.

Stell J. G. and Worboys M. F., 1997, 'The Algebraic Structure of Sets of Regions', in S. C. Hirtle and A. U. Frank (eds.), *Spatial Information Theory: A Theoretical Basis for GIS. Proceedings of the Fourth International Conference*, Berlin: Springer, 163–174.

Strawson P. F., 1959, *Individuals. An Essay in Descriptive Metaphysics*, London: Methuen.

Stroll A., 1979, 'Two Concepts of Surfaces', *Midwest Studies in Philosophy* 4, 277–291.

Stroll A., 1985, 'Faces', *Inquiry* 28, 177–194.

Stroll A., 1988, *Surfaces*, Minneapolis: University of Minnesota Press.

Sylvan R. and Hyde D., 1993, 'Ubiqutous Vagueness without Embarrassment', *Acta Analytica* 10, 7–29.

Talmy L., 1983, 'How Language Structures Space', in H. Pick and L. Acredolo (eds.), *Spatial Orientation: Theory, Research, and Application*, New York: Plenum, 225–282.

Tarski A., 1929, 'Les fondements de la géométrie des corps', *Księga Pamiątkowa Pierwszkego Polskiego Zjazdu Matematycznego*, suppl. to *Annales de la Société Polonaise de Mathématique* 7, 29–33 (Eng. trans. by J. H. Woodger, 'Foundations of the Geometry of Solids', in A. Tarski, *Logics, Semantics, Metamathematics. Papers from 1923 to 1938*, Oxford: Clarendon, 1956, 24–29).

Tarski A., 1935, 'Zur Grundlegung der Booleschen Algebra. I', *Fundamenta Mathematicae* 24, 177–198 (Eng. trans. by J. H. Woodger, 'On the Foundations of the Boolean Algebra', in A. Tarski, *Logics, Semantics, Metamathematics, Papers from 1923 to 1938*, Oxford: Clarendon, 1956, 320–341).

Tarski A., 1937, 'Appendix E', in J. E. Woodger, *The Axiomatic Method in Biology*, Cambridge: Cambridge University Press, 161–172.

Tarski A., 1959, 'What Is Elementary Geometry?', in L. Henkin, P. Suppes, and A. Tarski (eds.), *The Axiomatic Method*, Amsterdam: North-Holland, 16–29.

Taylor B., 1985, *Modes of Occurrence: Verbs, Adverbs and Events*, Oxford: Blackwell.

Taylor R., 1955, 'Spatial and Temporal Analogies and the Concept of Identity', *Journal of Philosophy* 52, 599–612.

Taylor R., 1959, 'Moving about in Time', *Philosophical Quarterly* 9, 289–301.

Thomson J. J., 1965, 'Time, Space, and Objects', *Mind* 74, 1–27.

Thomson J. J., 1977, *Acts and Other Events*, Ithaca (NY): Cornell University Press.

Thomson J. J., 1983, 'Parthood and Identity Across Time', *Journal of Philosophy* 80, 201–220.

Thomson J. J., 1998, 'The Statue and the Clay', *Noûs* 32, 149–173.

Thrower N. J. W., 1996, *Maps and Civilization*, Chicago: Chicago University Press.

Tiles J. E., 1981, *Things That Happen*, Aberdeen: Aberdeen University Press.

Tversky B., 1989, 'Parts, Partonomies, and Taxonomies', *Developmental Psychology* 25, 983–995.

Tversky B., 1997, 'Spatial Perspective in Descriptions', in P. Bloom, M. A. Peterson, L. Nadel, and M. F. Garrett (eds.), *Language and Space*, Cambridge (MA): MIT Press, 463–492.

Tversky B. and Hemenway K., 1984, 'Objects, Parts, and Categories', *Journal of Experimental Psychology: General* 113, 169–193.

Tye, M., 1989, *The Metaphysics of Mind*, Cambridge: Cambridge University Press.

Tye M., 1990, 'Vague Objects', *Mind* 99, 535–557.

Unger P., 1980, 'The Problem of the Many', *Midwest Studies in Philosophy* 6, 411–467.

Van Benthem J., 1983, *The Logic of Time*, Dordrecht: Kluwer (2nd ed. 1991).

Van Cleve J., 1986, 'Mereological Essentialism, Mereological Conjunctivism, and Identity Through Time', *Midwest Studies in Philosophy* 11, 141–156.

Vandeloise C., 1986, *L'espace en français: sémantique des prépositions spatiales*, Paris: Seuil (Eng. trans. by A. R. K. Bosch, *Spatial Prepositions. A Case Study from French*, Chicago: University of Chicago Press, 1991).

Vandeloise C., 1994, 'Methodology and Analyses of the Preposition *in*', *Cognitive Linguistics* 5, 157–184.

Vandeloise C., 1999, *Langue et Physique*, Stanford: CSLI Publications, forthcoming.

Van Inwagen P., 1981, 'The Doctrine of Arbitrary Undetached Parts', *Pacific Philosophical Quarterly* 62, 123–137.

Van Inwagen P., 1987, 'When Are Objects Parts?', *Philosophical Perspectives* 1, 21–47.

Van Inwagen P., 1990, *Material Beings*, Ithaca (NY): Cornell University Press.

Van Inwagen P., 1993, 'Naive Mereology, Admissible Valuations, and Other Matters', *Noûs* 27, 229–234.

Varzi A. C., 1994, 'On the Boundary Between Mereology and Topology', in R. Casati, B. Smith, and G. White (eds.), *Philosophy and the Cognitive Sciences. Proceedings of the 16th International Wittgenstein Symposium*, Vienna: Hölder-Pichler-Tempsky, 423–442.

Varzi A. C., 1996a, 'Reasoning about Space: The Hole Story', *Logic and Logical Philosophy* 4, 3–39.

Varzi A. C., 1996b, 'Parts, Wholes, and Part-Whole Relations: The Prospects of Mereotopology', *Data & Knowledge Engineering* 20, 259–286.

Varzi A. C., 1997, 'Boundaries, Continuity, and Contact', *Noûs* 31, 26–58.

Varzi A. C., 1998, 'Basic Problems of Mereotopology', in N. Guarino (ed.), *Formal Ontology in Information Systems*, Amsterdam: IOS Press, 29–38.

Vendler Z., 1957, 'Verbs and Times', *Philosophical Review* 66, 143–160.

Vieu L., 1991, *Sémantique des relations spatiales et inférences spatio-temporelles. Une contribution à l'étude des structures formelles de l'espace en Langage Naturel*, Université Paul Sabatier de Toulouse: PhD Thesis.

Vieu L., 1997, 'Spatial Representation and Reasoning in Artificial Intelligence', in O. Stock (ed.), *Spatial and Temporal Reasoning*, Dordrecht: Kluwer, 5–41.

Vieu L., 1993, 'A Logical Framework for Reasoning about Space', in A. U. Frank and I. Campari (eds.), *Spatial Information Theory: A Theoretical Basis for GIS. Proceedings of the Second International Conference*, Berlin: Springer, 25–35.

von Wright G. H., 1963, *Norm and Action. A Logical Inquiry*, London: Routledge & Kegan Paul.

von Wright G. H., 1979, 'A Modal Logic of Place', in E. Sosa (ed.), *The Philosophy of Nicholas Rescher*, Dordrecht: Reidel, 65–73.

Walker A. G., 1947, 'Durées et instants', *Revue Scientifique* 85, 131–134.

Weeks J. R., 1985, *The Shape of Space. How to Visualize Surfaces and Three-Dimensional Manifolds*, New York: Dekker.

White G., 1993, 'Mereology, Combinatorics, and Categories', Preliminary Report for the SNF Project 11.31211.91, Schaan.

Whitehead A. N., 1916, 'La théorie relationniste de l'espace', *Revue de Métaphysique et de Morale* 23, 423–454 (Eng. trans. by P. J. Hurley, 'The Relational Theory of Space', *Philosophy Research Archives* 5 (1979), 712–741).

Whitehead A. N., 1919, *An Enquiry Concerning the Principles of Human Knowledge*, Cambridge: Cambridge University Press.

Whitehead A. N., 1920, *The Concept of Nature*, Cambridge: Cambridge University Press.

Whitehead A. N., 1929, *Process and Reality. An Essay in Cosmology*, New York: Macmillan.

Wiggins D., 1968, 'On Being in the Same Place at the Same Time', *Philosophical Review* 77, 90–95.

Wiggins D., 1979, 'Mereological Essentialism: Asymmetrical Essential Dependence and the Nature of Continuants', *Grazer philosophische Studien* 7, 297–315.

Wiggins D., 1980, *Sameness and Substance*, Oxford: Blackwell.

Willard D., 1994, 'Mereological Essentialism Restricted', *Axiomathes* 5:1, 123–144.

Wilson N. L., 1955, 'Space, Time and Individuals', *Journal of Philosophy* 57, 589–598.

Winston M., Chaffin R., and Herrmann D., 1987, 'A Taxonomy of Part-Whole Relations', *Cognitive Science* 11, 417–444.

Wunderlich D., 1985, 'Raum, Zeit, und das Lexicon', in H. Schweizer (ed.), *Sprache und Raum*, Stuttgart: Metzler, 66–89.

Yablo S., 1987, 'Identity, Essence, and Indiscernibility', *Journal of Philosophy* 84, 293–314.

Yablo S., 1992, 'Cause and Essence', *Synthese* 93, 403–449.

Zadeh L., 1965, 'Fuzzy Sets', *Information and Control* 8, 338–353.

Zarycki M., 1927, 'Quelques notions fondamentales de l'Analysis Situs du point de vue de l'Algèbre de la Logique', *Fundamenta Mathematicae* 9, 3–15.

Zimmerman D. W., 1995, 'Theories of Masses and Problems of Constitution', *The Philosophical Review* 104, 53–110.

Zimmerman D. W., 1996a, 'Indivisible Parts and Extended Objects: Some Philosophical Episodes from Topology's Prehistory', *The Monist* 79, 148–180.

Zimmerman D. W., 1996b, 'Could Extended Objects Be Made Out of Simple Parts? An Argument for "Atomless Gunk"', *Philosophy and Phenomenological Research* 56, 1–29.

Zimmerman D. W., 1997, 'Coinciding Objects: Could a "Stuff Ontology" Help?', *Analysis* 57, 19–27.

Index

Symbols

WIN	whole location (inside)	142
WL	whole location	120
WOCφ	whole occupation relative to φ	135
WOUT	whole outside	143
−	difference	45
=	negative difference	149
−'	C-based difference	66
~	complement	45, 82, 150
+	sum	43
+	negative sum	149
+'	C-based sum	66
×	product	43
.	modal existence	155

Text

Theories